Project management of multiple projects and contracts

Edited by Jack Loftus

Thomas Telford

Published by Thomas Telford Publishing, Thomas Telford Limited, 1 Heron Quay,
London E14 4JD.
URL: http://www.t-telford.co.uk

Distributors for Thomas Telford books are
USA: ASCE Press, 1801 Alexander Bell Drive, Reston, VA 20191-4400
Japan: Maruzen Co. Ltd, Book Department, 310 Nihonbashi 2-chome, Chuo-ku, Tokyo 103
Australia: DA Books and Journals, 648 Whitehorse Road, Mitcham 3132, Victoria

First published 1999

A catalogue record for this book is available from the British Library

ISBN: 0 7277 2710 9

Typeset by MHL Typesetting Limited, Coventry
Printed and bound in Great Britain by MPG Books Ltd

Contents

Contributors

Jack Loftus BSc, CEng, MIChemE

Jack Loftus recently retired from the post of Central Engineering Manager, Thames Refinery, Tate & Lyle Sugars, responsible for all capital projects, planned maintenance and cleaning and all building works at the world's largest sugar refinery, located at Silvertown in the East End of London. He was with Esso Petroleum Co. Ltd from 1962 to 1973, dealing with process engineering, plant technical services, investment analysis, budget co-ordination and medium range planning. He then joined Total Oil GB Ltd, in turn as refinery Operations Officer, Project Manager for a refinery expansion, and Manager Refining Operations. He moved to Lummus Crest Ltd in 1985, as a Senior Project Manager and then was successively Manager of Project Engineering, a project director and Director of Projects. In 1987 he joined Glaxo Group Research Ltd as Manager Engineering Services, responsible for creating a new department to provide a full project and discipline engineering service to all the company's projects, and he became the Site Project Manager for the Stevenage Research Campus responsible for planning, organizing, directing and control of the Stevenage development. This included quality assurance, industrial relations, safety, and the achievement of cost and schedule targets for projects valued at £500 million. He is a regular speaker in short courses in project management run by the UMIST Project Management Group.

Dave Cole

Dave Cole has over 30 years experience in engineering design and project management in a chemical and food related process industry.

For the past 10 years he has held senior management positions, with overall responsibility for engineering design, project services and project execution, in a central engineering department accountable for a capital expenditure programme of multi-disciplined projects valued at £20–30 million per annum. Additionally he has had joint responsibility for the formulation and introduction of departmental engineering standards, specifications and project operation and reporting procedures, including procedures to ensure company compliance with the Construction (Design & Management) Regulations.

Jacqui Horne ACMA MAAT

Jacqui Horne started her professional life working for Kidsons Impey, Chartered Accountants. There followed a spell during which she taught at Mid Kent College, teaching AAT, Chartered Institute of Bankers, A Level Accounts and HNC/HND in Business and Finance courses.

She moved to Sight and Sound Education Ltd before spending time working in Europe for Quinton Hazell Ltd, a subsidiary of the US company, Echlin Corp., and gained experience working in Germany, France, Belgium, Holland, Italy and Spain.

Jacqui Horne moved to Railtrack in 1996 and is currently Financial Controller for the Midlands Zone, managing the finances of a £220 million portfolio of work.

David Major

David Major is a regional manager for a process contractor in the UK. Born in 1955, he gained qualifications in mechanical engineering, electrical engineering and building construction, and has taken part in and managed engineering projects all his working life. His first experiences were gained during school holidays at the age of 14 on a civil engineering project and he has subsequently worked on multi-disciplinary engineering projects in the food, process, oil, gold and diamond mining, effluent treatment, process, materials handling and water industries for contractor and client organizations in the UK, Spain, France, Nigeria, South Africa, Lesotho, Angola, Mozambique and Botswana.

Jim Rollinson ARICS MACostE

A quantity surveyor by profession with over 30 years experience of the commercial management of construction projects, Jim Rollinson has worked in a variety of different roles for clients, consultants, contractors and subcontractors. He has been involved in a wide variety of construction projects from small scale general building through to multi-million pound process plant. He is experienced in the full construction process including commercial strategies, cost planning, estimating, procurement, contract documentation, cost control, contractual procedures, account and claim negotiation.

Stephen Wearne BSc, PhD, CEng, FAPM, PMP

Stephen Wearne is a consultant and is a member of the Project Management Group at Manchester on problems of project and contract management. He was originally a factory apprentice and 'sandwich' student, and then worked on the construction, design, planning and co-ordination of water power projects in Spain, Scotland and South America. In 1957 he joined contractors on construction and design and then contract and project management of nuclear power projects in the UK and Japan.

In 1964 he moved into teaching and research on engineering management, first at the University of Manchester Institute of Science & Technology, and from 1973 to 1984 as Professor of Technological Management at the University of Bradford. His research has included design and project organizations, joint venture risks, engineering and construction contracts, project control, plant commissioning teams and the managerial tasks of engineers in their careers. He is author or joint author of the books *Control of Engineering Projects*, *Principles of Engineering Organisation* and *Engineering Project Management*. He was the first Chairman of the Institution's joint Engineering Project Management Forum and the Association for Project Managements Specific Interest Group on Contracts & Procurement.

1. The client's problem

J. Loftus

In all manufacturing companies there is an ongoing, continuous need for capital investment in projects which are required to meet changes imposed from outside the company. There will always be individuals, in all the functional departments, who will be bringing forward demands for new plant and equipment. Inevitably there will also be a constraint on expenditure imposed by the finance department, and the matching of cash supply to the perceived company needs will fall to that department charged with the execution of projects and contracts.

Anyone working in such a department will be aware, immediately, of the following problems:

- ☐ there are too many projects
- ☐ there is too little capital available
- ☐ there are too few resources to manage all the projects
- ☐ the definition of the projects is imprecise
- ☐ there is too much specified client preference
- ☐ there is a totally unrealistic appreciation of project costs
- ☐ the route to project authorization is not understood
- ☐ the time taken to reach authorization is not understood
- ☐ all projects must be undertaken at minimum financial risk to the company
- ☐ the department responsible for achieving contracts and projects is starved of resources
- ☐ the authority to approve expenditure is completely inappropriate to the task
- ☐ the existing workload within the capital projects department will be ignored by the person who authorizes the next project

From the above, it is essential that some order is brought into this anarchic situation, and that this order, or procedure, is agreed and accepted by the whole management structure of both the company and its individual manufacturing locations.

The procedure and system outlined here has been proven; it does work, it is not unique, but it is simple and may be applied by any organization.

Categorization and prioritization of projects

It is useful to put all projects into a number of simple categories, so that a clear picture can be presented to the decision makers and a consensus on priorities be reached quickly. The following has been found to be useful:

(a) projects greater than £250 000
(b) projects between £50 000 and £250 000
(c) projects less than £50 000.

These can then be sub-divided into 'need' categories as follows:

(1) essential (i.e. cannot be avoided or delayed)
(2) desirable (i.e. cannot be avoided but may be delayed)
(3) opportunity (i.e. can be avoided and delayed but offers an acceptable return on investment)
(4) other (i.e. anything else and projects under investigation).

A project listing can be created from the above, and used on an agenda for the setting of consensus priorities in each project cost category. It cannot be overemphasized that a high degree of discipline and rigour is required to ensure that only those really unavoidable projects are rated as essential. At this point it is worth pausing, because it is implicit in the above prioritization and categorization exercise that the total erected cost of each project is known or can be estimated.

This is, of course, not true. Those projects that have been adequately investigated and scoped can be estimated from either historical data or actual quotations. However, if the above exercise is done at the beginning of each financial year, the cost data on the majority of projects will be very imprecise. A number of 'short cut' methods are available, or can be developed from past experience. A good rule to use is to obtain budget quotations for the major pieces of equipment in each project and then use a multiplier of 4·5 to estimate the total erected cost, including all clients' costs before, during and after the project. This factor is remarkably robust for an industry where there is a mixture of solids handling and process plant. It will certainly provide an adequate screening basis for the above exercise.

Similarly, no detailed analysis of the savings of opportunity projects will have been performed at this screening stage. The best, simple, guide to opportunity project acceptability is whether a project has a crude return equal to or greater than the company's return on capital employed (ROCE) or return on net operating assets (RONOA). These returns are beloved by accountants and therefore readily available in all companies. At the screening stage, all that need be done is to see if the obvious savings generated by a project produce a return better than the above targets. A word of caution: achieved returns rarely meet those predicted at the screening stage, so a judgemental margin should be added. In some companies a safety factor of 2 is not unusual.

The exercise outlined above should be started 6 months before the beginning of the next financial year in order to define the capital funding required. It should then be refined at the start of the financial year in order to establish true priorities. This will produce an agreed, consensus workload for the department responsible for capital projects implementation. This workload will also define the resources required to perform the task, but these resources will depend upon the contract and project management strategy and philosophy employed.

Project and contract management

Capital expenditure and revenue expenditure are treated in different ways for taxation purposes. Generally, revenue expenditure is not depreciated and therefore cannot be used as a tax shield. Capital expenditure becomes part of the fixed assets of a company and can be depreciated and will form part of ROCE or RONOA. It is essential that the two types of expenditure are separate, and are managed in different, unrelated ways.

A method must exist to accurately and carefully monitor capital expenditure, which must be capable of finance audit both from inside and outside the company. This means that every action taken on a capital project should be recorded, so that the changes to the original authorized cost can be identified and justified.

The simplest method of imposing cost control, is to know the project cost accurately before authorization to spend the money is given. This imposes a considerable workload on the company wishing to make the investment, but the removal of uncertainty of the final cost also removes the risk of over-expenditure. It also minimizes the risk that the predicted return on investment will not be achieved because of project cost overruns.

It has been found to be successful to adopt the following simple procedure:

(a) for all projects greater than £10 000 in value, at least three competitive bids are mandatory
(b) bids should be invited on an adequate front-end package, consistent with a fixed price, fixed scope, fixed term tender
(c) the contract documents should be standardized and used on each suitable contract, with minimum 'customization'
(d) tender evaluation should follow a simple, but rigid, review procedure with the commercial and technical evaluation being performed separately and concurrently

(e) the project engineer, or project manager, who will manage the project should also be responsible for the production of the front-end design package and the management of the tender evaluation process

(f) the other members of the project team should participate in all the pre-award processes

(g) the whole of the front-end process should be planned and executed as a project in its own right. It should be cost controlled, planned and managed properly to agreed time, cost and technical objectives.

The department of the company responsible for capital projects is not an engineering department. It is a contract and project management organization, with only those technical resources necessary for it to perform its function. It is of paramount importance that it has in place those essential, standard documents and procedures to allow it to operate.

The most important standard document used in any capital project organization is the tender enquiry document. This 'set' of documents defines, or should define, quite clearly and unambiguously how the project is to be procured and managed. The standard document should only establish the minimum requirement for style and content, but it must never be significantly amended without the approval of the capital projects department manager. A successful tender enquiry document can comprise six principal sections, three relating to commercial matters and three to technical matters, for example:

☐ Instructions to tenderers and tender acknowledgement.
☐ Form of tender.
☐ Conditions of contract.
☐ Specification.
☐ Drawings.
☐ Schedules.

It is worthwhile to expand a little on each of these sections.

Instructions to tenderers

This section is mandatory to ensure a fair and regulated tendering process, it should state:

(a) submission date
(b) validity period
(c) control of site visits
(d) regulations for site contractors
(e) communication routes
(f) any matters peculiar to the specific project
(g) tender acknowledgement forms.

Form of tender

Whatever else a tenderer submits, he must complete a formal offer form sent out with the enquiry. This completed form, the scope and complexity of which may vary depending upon the type of project, constitutes the formal offer which may form the basis of contract for the works. Tenderers must not be allowed to substitute alternatives or modify the form.

Conditions of contract

It is essential that all significant project work is the subject of a formal contract. Tenderers must therefore be advised on the proposed conditions of contract in the enquiry document. Therefore, the capital projects department has to have a set of contracts, suitable for all works likely to be undertaken. These documents should be based on 'model forms' used in various fields and adapted for use by a particular company. The company's legal advisers should be consulted and participate in the production of these standard contracts. If this is done it will avoid the need to refer each contract back to the lawyers, unless a major change is needed which cannot be negotiated out during the final discussion with the preferred contractor.

It is worth noting that in 3 years of capital project management, covering some 300 projects, it has been possible to use these forms, without major modification, to the satisfaction of both client and contractor.

Specification

It is desirable to standardize specifications and associated schedules, to minimize effort and avoid re-inventing the wheel for each project. It is also necessary to ensure that the minimum quality is defined in order to ensure that the tender responses are suitable, in both content and value for money, to give good project cost control. It has been found by experience that the following content is desirable:

(1) General: introduction, statement of technical requirements, project description, project location, project works to be undertaken, project timing, site conditions, working arrangements, safety, daywork overtime, provisional sums.
(2) Scope of the works: scope of design, deliverable information to the client, enabling works, battery limits, building works, plant and equipment, mechanical services, electrical services, instrumentation, 'free issue' items, project management, options, exclusions.
(3) Design – the scope must be clearly defined and the design responsibility clearly and unambiguously stated. Also, the level of design input required must be consistent with the standard contract forms used. Basic requirements: process basic principles, feedstock, plant and

equipment, services, buildings, standards, statutory requirements, existing plants/structures/services, existing data, contractors' data.

(4) Materials and workmanship: general – reference to British Standards, codes of practice, regulations etc.; particular – company-specific specifications and standards. For example, electrical installation, mechanical handling equipment, conveyors, fire and safety, dust explosion prevention and protection aid, etc.

Reference should also be made to quality and quality assurance in this section. Also suppliers and subcontractors should be specified but only where no alternative is available or desirable.

The quality and content of programmes and progress reports should be stated, as should the types, sizes and scales of drawings. It is very important to differentiate between drawings and documents for 'approval' and those for 'review and comment'. At the limit, the client should not want to approve any contractor produced documents.

Other topics to be covered are manuals, spare lists, lubrication schedules, testing and commissioning, definition of testing, takeovers, precommissioning, commissioning, performance tests, acceptance, and finally training.

The above may appear to be a horrendous task, but once performed it can be heavily plagiarized for most succeeding projects. A set of model specification clauses will greatly speed up the work of producing a specification.

Drawings

All available drawn information should be provided in the tender enquiry and listed in this section for the document.

Schedules

The purpose of the schedules is to incorporate technical details of a supplementary nature with the enquiry document without overburdening the specification. Schedules may relate to

☐ standard documents
☐ procedural notes
☐ statistical data
☐ other relevant data sources.

Contractor selection

Having decided what you want to do and how it is to be done, the next, and in some ways the most difficult, problem is to find someone to do it. There exists a whole world of contractors out there, willing and apparently able to do anything you request for suitable reimbursement.

It is essential, for all projects with a value of £1 million or greater, to compile a list of potential contractors, and to send to them a simple questionnaire to see if they want to be considered for work at your location. All the contractors on the list will say 'yes'. Having done the necessary financial checks and sought recommendations from their nominated (and satisfied) clients, produce a short-list of up to six contractors and go and visit them at their head offices. A checklist should have been prepared and a scoring system established so that a measure of suitability can be quickly ascertained. The most important question that a client can ask a contractor is 'What is the minimum, total erected cost contract, for which your organization considers it can compile a competitive bid?'.

The obvious supplementary question is the name of the last client at the cost level. Contractors must be honest, and if their reply puts the minimum size of job out of your league, they should be courteously eliminated. It is important that you match the contractor's capability and experience to your job. If the contractor is too big, his overhead structure will make his price uncompetitive. If he is too small then it is unlikely that he can manage the job. Choose carefully, so that the risk of the contractor failing or treating the client as an also ran, is minimized.

A contractors register, compiled from each bidding competition and completed project, is the best reference list a client can produce. A considerable effort must be made to match the contractor to the job in terms of value, experience and expertise.

It should be noted that the use of the word 'contractor' in this context is quite specific. It does not include consultants, consulting engineers or vendors. It is used in the sense that the 'contractor will provide a full scope service, of detailed design, procurement and construction'. The consultant, consulting engineer and vendor can only provide part of this service and can, of course, be so used by a contractor or indeed by a client.

Use of consultants

In the management of small projects and contracts, consultants are used to provide highly specialized knowledge in those areas where the client cannot afford to retain such expertise because of its infrequent use. There will always be in-house, highly specialized knowledge about the business of the company. However, it is unlikely that specific knowledge will exist, for example, in the following fields:

☐ corrosion
☐ metallurgy

- [ ] electrical engineering
- [ ] strength of materials
- [ ] combustion technology
- [ ] non-destructive testing
- [ ] civil and structural design.

There are many organizations able to meet a client's needs, and suitable consultancy agreements can be reached either on an *ad hoc* basis or as an on call service. A standard consultancy agreement is recommended, and frequent checks on the competitiveness of the consultants currently employed are always necessary.

Use of consulting engineers

The main use of consulting engineers is to supplement the client's organization when it is resource limited. However, the usefulness of such assistance will depend upon the quality of the brief and the client's management available to manage the consulting engineer. If you are tempted to use a consulting engineer, it is recommended that a bidding competition is set up as if the consulting engineers were, in fact, merely contractors. This should ensure better value for money. In practice, it will always give the best value if the consulting engineer is brought in at the earliest possible stage, and he is employed to produce finite, limited pieces of work. At each stage the quality of his output can be assessed before further commitments are made, for example, the stages could be those in the ACE and RIBA agreements. It is important that financial and schedule targets are agreed before the work begins and that it is effectively monitored. Consulting engineers' costs will be high compared to the use of in-house resources and must be effectively controlled, especially in the early stages of a project where definition is less precise.

It is unlikely that an organization involved in the management of small projects and contracts will ever use a consulting engineer as a project manager because of the cost involved. One service offered by consulting engineers which can be of great value is technical and project auditing.

Use of vendors

Suppliers of equipment will usually offer an installation service. However, the limits of supply of such a service need to be carefully examined and understood. The listed exclusions are usually vague, but in practice cover most civil engineering projects, buildings, power, utilities, infrastructure, access, site conditions, commissioning and testing. The message is usually that the vendor will site erect his own equipment but someone else has to provide all the essential supplies and connections to make it work. The client therefore has a choice:

(a) For simple applications appoint contractors, managed by the client, to provide the service required by the vendor and specified by him in advance.

(b) For complicated applications, appoint a main contractor to manage both his own, his contractors and the vendor's installation personnel.

The definition of 'simple' and 'complicated' may appear too vague. A rule of thumb is that if the vendor-supplied equipment is a straightforward purchase requiring no engineering or design of services then use route (a) above. If design of services is needed, then use route (b). There is an obvious intermediate case which is dependent upon the availability of in-house resources, that is, the client can also act as main contractor.

The department – project organizations

Having defined the task and the method of working to achieve the end result, it is necessary to define an organization capable of performing the task.

The organization described below was created with the following objectives:

(a) to manage all capital projects at two production sites based in the UK
(b) to manage an annual capital expenditure of £20 million
(c) to manage a maximum of 10 projects with a total installed cost of more than £500 000 per project
(d) to manage a maximum of five projects with a total installed cost greater than £2·5 million per project
(e) to manage a total of 90 projects through to completion in each financial year
(f) to manage the investigations through to authorization of 70 projects in each financial year
(g) to produce a monthly progress report for each project
(h) to manage the prioritization of projects for the next financial year
(i) to perform a full cost management function for all investigations and projects in progress
(j) to provide a full planning service to all projects and investigations in progress
(k) to maintain and administer the engineering records of all existing plant and machinery
(l) to produce follow-up reports on each completed project
(m) to participate with external auditors in the technical and project auditing of all projects and investigations estimated to have a total installed cost greater than £500 000

(n) to establish the training needs, both technical and personal, for all members of the department

(o) to optimize the use of available resources to give the best value for money project management to the company

(p) to support and advise all other company departments on project and contract management

(q) to produce departmental procedures for all project and contract management activities

(r) to produce engineering standards, specifications and codes of practice

(s) to provide a full secretarial service to all project managers and project leaders

(t) to provide a photographic service to all project managers, project leaders and all company departments.

The departmental organization designed to fulfil these objectives has 40 positions, each covered by a written job description. There are four functional groups reporting to the capital projects manager. Currently there are 18 company employees and 22 agency and contract staff. The department is designed for a company engaged in a 'low technology, warm and wet' industry and so can afford to be lean on technology that is:

☐ highly specialized
☐ high temperature
☐ high pressure.

The capital projects department functions as a matrix with four operating groups. These are described below.

Project services

This group provides a complete project service for all projects, both large and small. It also acts as a central administration unit for all project information for reporting progress and cost. It provides the following functions on a matrix basis to all project managers, project leaders and project investigation leaders:

☐ planning
☐ progress measurement
☐ commitments reporting
☐ expenditure reporting
☐ cost forecasting
☐ progress photography
☐ estimating
☐ typing.

The project services section also provides the capital project accounting and capital project financial reporting function for the division and the company. Other services provided are engineering records and archiving, photography and an insurance and risk management service. The manager of the project services group also acts as project manager for a major project, but only one at a time.

Project engineering management

This group is managed by a project engineering manager, who essentially is a manager of project leaders for all projects with a value of up to £5 million.

The group also contains the required in-house discipline engineering resource to meet all project demands. The group is split into five operating areas:

- mechanical
- civil and structural
- instrumentation and electrical
- project engineering
- contract services.

The functions are self-explanatory, except contract services. This small group ensures that all project and contract procedures are followed and that standard forms of contract, etc., are always used on all projects. They also produce payment certificates and administer all claims and valuations. This ensures that a common approach is taken on all projects and that a meticulous central record is kept of all contracts, changes, variations, bids, bid evaluations and contractual negotiations. This ensures that the departmental manager is kept fully aware at all times of the position on all projects and also that the lessons learned on one contract are incorporated into the management of the next and all future contracts. The contract services group also keep an up-to-date register of all bids requested, bids received, bid evaluations, contracts awarded and contractor performance. They are located within project engineering management rather than project services because of the day-to-day interaction necessary with the project engineers.

Technical liaison

All projects go through an investigation stage, and are frequently led, in the early stages, by personnel outside the capital projects department.

There is an obvious need for planning, resourcing, and definition of the quality of information to be passed to the capital projects department so that authorization and contract execution can proceed to a realistic timetable. This diplomatic, co-ordination function is provided by one experienced project manager as a sideline, while he manages his own, large project.

Project management

There will always be projects which are of particular importance to the company because of cost, technological change, urgency or confidentiality, that need a dedicated project manager. Finding the right individual and retaining such an individual within a small project and contract management organization is a challenge for the manager of the capital projects department.

The level of permanent staff within the department and the decision to use external resources is always based on the company's perception of the longer-term need for capital project management. A 'core' team will always be needed, even in those periods when capital expenditure is restricted to only the barest of essentials.

There are certain functions which cannot be provided easily at the correct level of professionalism and experience within an operating company's salary structure, for example cost management and planning. Similarly, it is prudent to use a civil and structural practice to provide a service backed by professional indemnity insurance. Such a service will also bring a deep knowledge of the local planning and building regulations.

It has been recognized that using dedicated CAD designers to do the majority of the detailed drawing work, based on outlines produced by the engineers, is quicker and much more cost-effective than using the engineers to produce the final drawings. This is almost certainly a function of the workload within the department, and the fact that most engineers are working on at least six projects at any one time. It also has the advantage that the defined quality of drawings produced in-house as part of the front-end package for bidding purposes is always consistent because it is done by the same people. Good CAD designers are also hard to find, but if the right CAD tool for the job and a variety of interesting and motivating work is available, they do stay for long periods.

Project audits

It is excellent practice to employ a disinterested, experienced third party to perform technical and project audits on all major projects.

This is not the statement of a masochist but is based on experience. The only way to avoid being the victim of your own systems and procedures is to open them up to the review and comment of others. Also to achieve timely authorization of major projects, the unbiased support of the auditor will carry far more weight than the well assembled and presented arguments of any operating department.

There are dangers in an auditor becoming, over a long period, too familiar with the department he is auditing, and tending to audit the people and not the project and procedures. The auditor should be changed frequently and only appointed for limited periods. The appointment should also be to tight terms of performance; a suggestion of such terms is included in a later chapter.

2. Capital expenditure proposals

D. Cole

Introduction

Chapter 1 identifies and describes a client's problems in relation to his ability to formulate, define, resource and manage implementation of a capital expenditure ('Capex') programme.

This chapter endeavours to demonstrate how the client should, with the aid of project management expertise using established practices, develop a project from inception through to provision of a formal Capex proposal and its submission for authorization.

It is acknowledged that individual projects will require particular skills and methods of application. However, it is anticipated that consistency in approach and uniformity in procedures should be employed in project investigation and execution phases.

Overriding responsibility for strategy and corporate policy on investment of capital must rest with the client. The project management role lies in developing such strategies to enable successful implementation of the client's capital expenditure plan.

Client responsibility and liaison

Responsibility

A company strategic plan for capital expenditure will have been or need to be formulated in recognition of corporate objectives and future business vision.

Directorate strategies, company directives, research and development, investment opportunities and legislative requirements are incorporated into a business strategy which forms the basis for the strategic plan for capital expenditure.

Identification of proposed projects, their first pass justification, prioritization and cost/time estimates enables compilation of a Capex strategic plan.

Versions of the strategic plan cover periods of, say, 5 or 3 years, the first year of which will be developed into a detailed annual operating plan for capital expenditure.

Responsibility for all strategic planning including the setting and authorization of the annual operating plan belongs to the client. Subsequent review or authorization to revise the annual operating plan during project proposal development or project execution must be the client's responsibility.

Signing for authorization of capital expenditure requests that trigger release of funds for project implementation again is the sole responsibility of the client.

The project manager's role in the strategic planning phase is limited, assisting where necessary on methodology and guidance on provision of cost and timing information.

Liaison

The project manager or his appointed investigation leader is responsible for establishing the project strategy, development of design, investigation, cost and timing estimation, preparation of formal project proposals and request for authorization documentation.

During these phases close liaison with the client and representatives of the appropriate business areas is essential. An agreed project strategy needs to be established early. Routes for design, commercial practice, investigation methods, provision of financial justification and project proposal approval may be subject to client policy and in-house resource availability.

Key to the successful execution of a project is the ability to provide a full, inclusive brief and to develop it into a comprehensive project proposal.

The client will be responsible for the provision of any specialist business or process knowledge vital to the development of the design brief and project proposal specification. Getting the project scope correct 'up front' and reducing the risk associated with late changes or additions can only be achieved by close project/client liaison.

Progress of both individual projects and investigations and of the annual operating plan need to be monitored. Formation of a capital review group representing the client, his business area representatives and the project management function can be used as a monitoring forum. A meeting agenda would include, investigation and project progress reports, review of annual operating plan status, revisions/additions to the annual operating plan (AOP) and prioritization of new Capex requests for proposals identified after the annual operating plan was established.

During project implementation client liaison needs to continue. The level of client involvement in procurement in terms of authorization of orders and expediting will be determined by his own policy. Approval of detail design and engineering is another area likely to attract client interest. Other areas of client involvement, dealt with in more detail in Chapter 5, needing close liaison are training, commissioning and testing, plant takeover and acceptance.

Project management policies.

Services to be provided to the client by a project management function are best defined by a set of objectives arranged in the form of project management policies.

These objectives should clearly set guidelines on the degree of project management support and responsibility for the various aspects of capital expenditure planning, project development and execution.

Policies

☐ To give guidance to the client in methods of formulating and maintaining a strategic capital expenditure plan to meet his corporate objectives.

☐ To assist the client in the identification of projects and viable proposals to enable development of a Capex annual operating plan.

☐ To give direction on definition, planning, costing and preparation for authorization of identified projects and for their management with minimum risk through to implementation within agreed performance targets, time scales and authorized budgets.

☐ To develop, establish and maintain controls and procedures to ensure safe, smooth running and successful implementation of all capital expenditure projects.

☐ To provide accurate and timely reporting of progress, financial status and to maintain effective project administration with regard to commercial and contractual matters.

Methods and principles for achieving the outlined objectives are fully expanded in the relevant sections of this chapter for pre-authorization activities and in Chapter 5 for project implementation.

Project brief

The basis for achieving project success lies in the provision of a comprehensive and timely project brief. The importance of clear identification of project scope, process description, definition of specific plant and process requirements, outline plant and material specifications and any statutory directives is paramount.

The nature and depth of the project brief will be determined by the size, value and complexity of the proposed project. Only if every effort is made to 'get it right' during the initial phase can later problems, associated with change of scope and additions to specification with their financial and timing implications, be alleviated.

Quality of the scope identified in the project brief is essential for early development of the overall project strategy in terms of requirements for investigation, design, procurement and implementation methodology.

A project brief will be initiated by identification of a capital expenditure need and developed through 'first pass' feasibility study to become the basis for establishment of a project proposal by way of a formal Capex request document. Acceptance and registration of the Capex request will trigger commencement of a full investigation and development from the project brief of a detailed design brief.

The Capex request process and project/development are described in detail below.

Project phases

All projects undertaken on behalf of a client will be subject generally to procedures for capital expenditure (Fig. 1). The client's Capex procedures may currently be in place or will need to be adopted in line with those defined in particular chapters of this book.

These procedures will basically dictate the sequence of events from project conception to completion and determine the various project phases as a series of business processes.

Capex business processes

Capex strategic planning
The process for formulation of a company strategic plan for capital expenditure is detailed in Chapter 1 and is normally the responsibility of the client.

Capex annual operating plan
The Capex strategic plan process will have identified a proposed project list, estimated 'first pass' costs and timing, tested the scope and justification against the business plan and provided a plan for capital investment for a predetermined time period.

The first year of such a plan is nominated as the Capex annual operating plan (AOP). An AOP will set monetary and timing targets and set priorities for project investigation, authorization and implementation.

Capex request process
Projects identified and prioritized to proceed within the AOP are processed through a formal Capex request documented system.

The process follows a route through the identification of specific Capex need by a business area; a feasibility study of the proposal, cost and timing; presentation of a proposal in standard format; further prioritization of

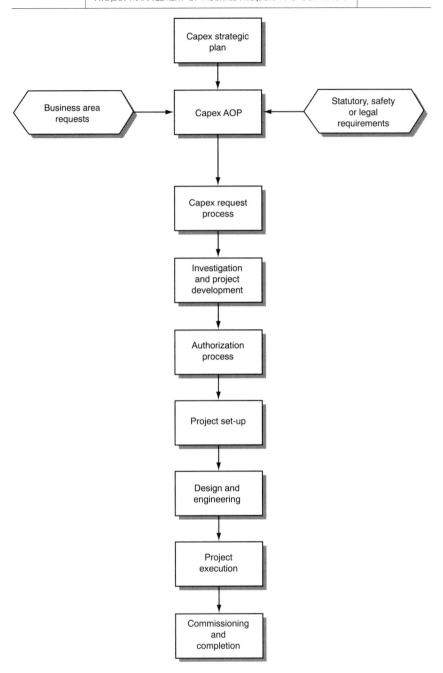

Fig. 1. Capex project phases

requests and decision on progress of a project to the investigation and development phase.

Investigation and project development

The objective of the investigation process is to establish the viability of a project proposal. All investigations require development of a project strategy, design brief, preliminary programme, cost plan estimate, project proposal and an outline procedure document.

The investigation will culminate in the presentation of a formal proposal assembling all investigation data in a format necessary for proceeding to requests for authorization.

Authorization

The purpose of the authorization process is to establish a record of formal submission of the assembled project investigation data in a standardized format for authorization at the appropriate authority level.

Set-up project

The process of setting up a project establishes the project manager's tasks and responsibilities for execution of the project.

Areas that need to be addressed during this phase are confirmation of project, commercial and contractual strategies; procurement, cost control and planning procedures; project administration and reporting.

Design and engineering

Confirmation of the project design strategy will facilitate the development of conceptual or front-end design into detail design for engineering and manufacture.

Responsibility for design, establishment of design criteria, safety and legislative design obligations and procedures for monitoring design are all areas to be considered.

Project execution

The project execution phase consists of procurement, manufacture, expediting, construction, installation and testing activities.

Establishment of procedures for monitoring progress, costs and quality of the various elements against the project programme, budget and specification is a prime project management responsibility.

Commissioning and completion

The management of commissioning demands early identification of scope, definition, procedures, programme, schedules, resource and training requirements and will involve close liaison with the client.

Project completion is largely a contractual process involving a sequence of activities leading to project acceptance by the client.

Procedures for the Capex strategic planning and Capex annual operating plan processes are expanded and detailed in Chapter 1.

Pre-authorization procedures including the Capex request process, investigation and project development and authorization process are described in more detail below.

Procedures for project implementation phases, project set-up, design and engineering, project execution and commissioning/completion are contained in Chapter 5.

Capex request process

Projects identified and prioritized within the annual operating plan need to be processed and documented in a formal and auditable system. A description of a sequence of events, administration of associated documentation and monitoring of progress is set out below (Fig. 2).

Sources of requests

The Capex request process begins with the identification of a specific need for expending capital and may be instigated by any of the client's business areas. Generally project proposals will be developed from the annual operating plan identified needs but allowance must be made within the plan for investment opportunities identified, as essential to the company, after the AOP has been set.

Feasibility study

Details of the request need to be developed in sufficient detail to enable the viability of a proposal to be established. Proposals will be progressed by way of a feasibility study where project scope and description, specific process requirements, outline cost and timing estimates, cost benefits, justification and consideration of alternatives can be examined. The main purpose of this stage is to present a proposal that can be demonstrated to warrant more exhaustive investigation leading to project authorization that can be justified in financial terms.

Generally, feasibility studies will be led by the client business area responsible for the Capex request as technical know-how and expertise are normally available in this area. The project management function will act in a support role providing a management, cost estimating and planning service.

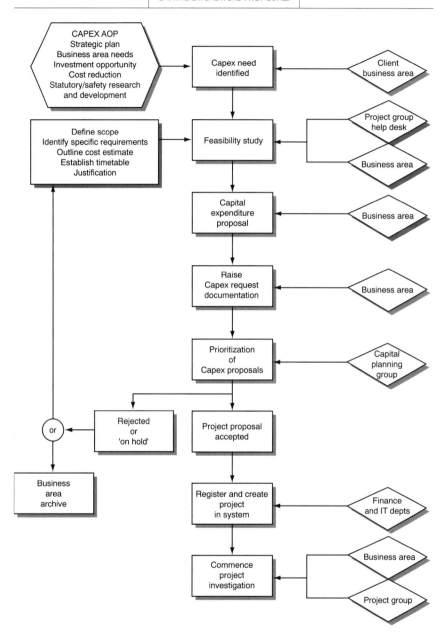

Fig. 2. Capex request process

Capex request documentation

Data derived from the feasibility study need to be assembled as a formal proposal and be presented in standardized format on a Capex request form. The form will need to take account of the complete authorization process and be designed for both proposal summary and request for authorization. A sample form is given in 'Pre-authorization reporting and documentation', below (see Fig. 8).

Part A of the document is headed 'proposal summary' and sets out, in detail appropriate to the project size/value, the proposed scheme. The form contains sections to be completed for project scope and description of proposal; objectives to be achieved by the project; expected benefits and financial justification; estimated costs for both capital and expense with claimed limits of accuracy; proposed time-scale for authorization and project completion; allocation of investigation leader and signatures for approval for Capex request.

Completion of the form is the responsibility of the appropriate client business area who should nominate a person as the document originator. Approval and signature of the request document prior to submission is by the relevant business area manager.

Part B of the Capex request form is concerned with the request for project authorization and is described in 'Authorization process', below.

Monitoring and control

Planning, monitoring and control of capital expenditure can be achieved by careful audit of the Capex annual operating plan and Capex requests by way of the services of a capital planning group. The purpose of a capital planning group is to scrutinize on a regular basis the progress of ongoing projects and investigations, the current status of the annual operating plan and all submitted Capex request proposals. Personnel representing each of the clients business areas and the project management function should form the group, which would be chaired by a nominee of the client. Planning group meetings should be held on a regular basis with minimum monthly intervals and be managed to an agreed agenda.

Review of submitted Capex request proposals would form a major part of a planning groups responsibility. Consensus across the group would be sought on the value of claimed benefits and justification of a proposal. Proposed projects included in the AOP would have cost estimates checked against the planned values. Non-AOP projects would need to be matched against available funds and justify their addition to the AOP. Decisions available to the planning group are:

☐ Acceptance of proposal and proceed to investigation phase.
☐ Reject the proposal as not viable.

☐ Place proposal on hold to await availability of funds.
☐ Review proposal scope and costs and re-submit.

Prioritization of proposals going forward for investigation will be made against availability of investigation resources and timing implications on the annual operating plan.

Summary progress reports on live projects and investigations would be reviewed and financial forecasts examined for identified variances with the annual operating plan. Any significant increase to project forecast values over planned values, that could be justified but not contained within the project contingency, would require revision to the AOP. Increased forecasts costs for investigations would also require AOP revision unless an alternative cost effective proposal or reduction in project scope and cost could be achieved.

Revision to the AOP and subsequent issue of a revised annual operating plan (RAOP) is the prerogative of the client. The capital planning group will review the plan and identify where revisions or additions are required and advise the client on how such changes would impact on the overall capital expenditure programme.

Project creation and registration

Following discussion and acceptance by the capital planning group of a Capex proposal the capital request form is registered and a unique CR number is allocated. This number identifies the request from registration through investigation to authorization and during project implementation.

Allocation of a CR number creates the project in the system and following consideration of resource availability and appointment of a project leader the project investigation phase commences.

Investigation and project development

When a potential capital expenditure project has been identified and registered, its accompanying Capex request document provides a basic description of the proposal. The objectives of the investigation and project development phase is to establish the viability of the proposed project. All investigations require the development of a project strategy, a level of front-end design, cost estimate, preliminary programme and an outline project procedure (Fig. 3).

The investigation will culminate with the assembly of all investigation data into a formal proposal presented in a format that meets the requirements of the authorization process.

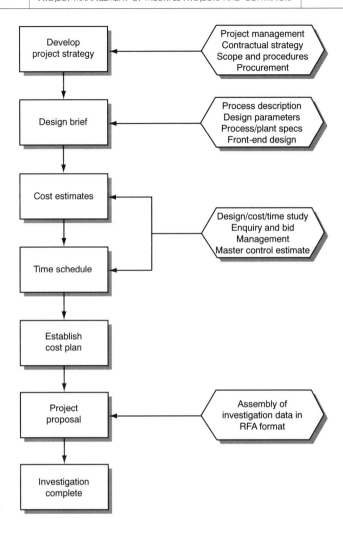

Fig. 3. Investigation and project development

Strategy

At the earliest opportunity in an investigation a comprehensive implementation strategy should be developed. The strategy incorporates methods of investigation necessary to provide a viable proposition to submit for authorization, that is, establishing design, timescale, cost plan and commercial arrangements; and anticipated methods of implementation – procurement, manufacture, construction and project administration.

Strategy selection will need to take account of several interrelated factors each of which can be influenced by the size and complexity of the project:

- [ ] Resources with appropriate skills to meet project needs.
- [ ] Method of design implementation.
- [ ] Time frame imposed by AOP and client operational requirements.
- [ ] Proposed contracting/commercial arrangements.
- [ ] Administration and management.

Having established an agreed project strategy, the project leader should prepare an investigation/project plan confirming the basic details of the strategy. The plan can be provided in flow chart format listing plan activity, activity description and responsibilities (see Fig. 5). A summary bar chart (see Fig. 6) setting out the proposed timetable should accompany the plan.

The project leader should also prepare a draft scope and procedures document for the project, the document being finalized and issued at commencement of project implementation (see Fig. 7).

Design

Design is likely to be a significant element within any project. It is essential that full consideration be given to the method of design implementation at the various phases of a project.

Selection of a method of design and the associated responsibility will be related to the following factors:

- [ ] Clients policy on control over key elements of process design.
- [ ] Availability of technical skills to develop design 'in-house'.
- [ ] Level of design/specification required prior to authorization.
- [ ] Use of consultants or contractors for design.

A review of these factors and their possible impact on the overall project strategy will determine the degree of allocation of design responsibility to the project team, consultants or contractors.

Limited availability of client 'in-house' resources can be overcome by elements of project design being undertaken by consultants. The term 'consultant' can be applied to professional or technical organizations as well as vendors or contractors acting in a similar capacity. Consultants may be employed as specialists in a particular design discipline or in a general role to carry out a full project design and cost study.

Investigation phase design is normally comprised of three principal elements – process, conceptual and front-end.

Process design is defined as the technology peculiar to the client's business and expertise is most likely to be found within the client's resources.

Conceptual design is the basic outline design of the project proposal and is developed from data submitted with the capital request.

The capital request form will contain basic details of plant and process specific requirements gleaned from the feasibility study carried out during the Capex request process.

Criteria necessary to satisfy process and conceptual design are encapsulated in a formal document named as the project design brief.

Front-end design is defined as the development of the design brief into a finalized workable proposal against which enquiries for fixed prices from vendors and contractor can be placed.

The project will be engineered at the front-end stage to establish the working criteria and to confirm process and engineering performance.

Dependent on project size and complexity front-end design may require consideration of

- confirmation of client requirements
- site survey and investigation
- preparation of process and instrumentation diagrams (P & IDs) and process control philosophy
- preparation of plant layouts and data sheets
- preparation outline civil/building plans and elevations
- preparation of detailed specification/scope of works
- preparation of technical elements of enquiry documents.

During the front-end stage full consideration is given to all aspects of project safety and shall relate to design, construction, plant operation and maintenance. Requirements of construction safety legislation contained within the Construction (Design and Management) Regulations are to be studied.

The Construction (Design and Management) Regulations came into force in 1995 and are applicable to project works involving engineering or building construction. The Regulations relate to construction safety and extend responsibilities beyond the contractor to designers and to the client.

There are clear requirements for the adoption of formal procedures and documentation by all parties to the construction process.

New specific duties are placed on the client and the contractor to ensure that health and safety is addressed, co-ordinated and managed through all phases of the project.

Additional roles have been created and the function and responsibility of such roles identified. The project phases have been extended beyond execution to include any subsequent maintenance, repair and demolition activities.

The regulations require the preparation and development of two documents for each project, the *Health and Safety Plan* and the *Health and Safety File.*

The Health and Safety Plan draws together safety information relating to the proposed project site, design and construction methodology. The plan is prepared during the design/investigation stage and developed as a working document during the engineering and construction phases.

The Health and Safety File is, in effect, a maintenance manual that brings together all the technical and construction data relating to the project. The file has the added purpose of identifying health and safety risks associated with the maintenance and removal of plant and structures.

Design brief

Completion of a Capex request form will have determined the basic description of a project proposal and recognized any specific process and plant requirements. Capital planning group acceptance of a proposal signals progress to investigation stage and prompts the need to develop a design to identify and detail all process and conceptual design requirements.

This design will be developed and captured in a design brief. Full project front-end design will be based upon the design brief and provide the project scope presented for authorization. The design brief is set out in formal documents and appended to the Capex request form at appropriate stages during investigation development.

Appendix 1 is a declaration, completed prior to the design brief presentation, that the deliverables accompanying the brief are accepted by the signatories as the design criteria for the project.

The form is signed by appropriate levels of the client's management structure and given authority to proceed to front-end design. The form also contains notes setting out responsibility for

☐ obtaining design brief approvals and signatures
☐ preparation of deliverables and their collation
☐ progress of design brief to front-end design.

Appendix 2 sets out a comprehensive list of design brief topics that need to be addressed during a design study. Each project will need to be assessed against the list to identify relevant elements. The form provides columns for identification of responsibility for preparation and approval of deliverables against each item.

Appendix 3 is a form designed to capture the approval of process design and project scope contained in the finalized project proposal. The form is to be completed at the finalization of front end design and appended to a subsequent request for project authorization. Contained in the form is a declaration and signatures of acceptance that the proposal submitted conforms to the requirements of the project design brief.

Sample forms for Appendices 1, 2 and 3 are given in 'Pre-authorization reporting and documentation', below (see Fig. 10).

Cost and timing

An objective of the investigation process is to define and establish the estimated cost and time-scale of a project proposal.

During front-end design a process and engineering design will have been established. Drawings, specifications and other relevant deliverables would have been assembled into a proposal on which 'in-house' or externally sourced cost estimates could be based.

In-house cost estimates are a compilation of cost allowances against the various elements of the proposal and cover the full scope of project implementation. Cost allowances for the investigation, project management during implementation and a project contingency to reflect the accuracy of the estimate would be included.

Externally sourced prices are obtained by an enquiry process which relates to the procurement of goods and services from suppliers, vendors or contractors.

A full description and explanation of the enquiry, bid management and appraisal process is contained in Chapter 3.

An outline project programme is to be provided at a level appropriate to the size and complexity of the project proposal. Analysis of the principle activities to determine their sequence, interdependence and time duration, enables networks, bar charts and schedules to be developed and key milestone dates to be identified. Expansion of planning procedures and methodology are described in 'Project planning' in Chapter 5.

Project cost plan and budget

The provision of a cost plan is a necessary ingredient in the investigation of a project and is essential in establishing the economic viability for authorization. Post-authorization the cost plan becomes the project budget and provides a yardstick for cost control as the project works proceed.

The format in which the cost plan is presented for inclusion in the authorization document matches the setting out of the project budget which is developed for cost control by way of the accounting system. The format is comprised of three principle elements – on-site costs; off-site costs and contingency.

The template for a master control estimate follows this format as it is developed through investigation to authorization and then monitored against actual expenditure during project implementation. A sample template is contained in 'Pre-authorization reporting and documentation', below (see Fig. 11).

On-site costs are defined as all material and labour costs associated with plant, equipment, services and structures required for a project, including manufacture, installation and commissioning.

Off-site costs will capture all the client and administration costs mainly accounted for by project management costs and costs incurred by the investigation process.

Contingency enables an allowance to be made within the plan for costs arising from unforeseen eventualities within project implementation and to reflect the degree of accuracy of the cost estimate.

The project cash flow plan is developed from and is dependent for accuracy on the integrity of the programme milestone dates and of an effective budget breakdown achieved by way of the master control estimate.

Project proposal

The project proposal is the culmination of project investigation and requires the collection of the assembled data into a recognized format. The format is set to reflect 'Request for authorization' documentation.

Responsibility for compilation of the project proposal lies with the project leader and for subsequent completion of a request for authorization he will have reviewed the following elements:

☐ scope – definition of the nature, extent and objectives of the proposal
☐ confirmation that the original investigation brief has been achieved
☐ strategy – identification of the means of executing the project
☐ design – technical development and specification of a viable working scheme
☐ timetable – establishment of an achievable programme for project implementation
☐ cost – confirmation of economic viability and establishment of a project budget
☐ justification – achievement of the technical and financial objectives of the project
☐ alternatives – consideration of alternative proposals and options.

All project investigations may be subject to technical and financial audit. The project leader must ensure availability of files containing all information relevant to the investigation and data in support of the 'request for authorization' proposal.

Authorization process

It is essential that the final proposal for a capital expenditure project is assembled in standardized formats. The purpose of this process is to clearly document the pertinent details in summary forms and to present them as a *Request for Authorization*. The need to submit a formal request for authorization will form an integral part of a client's company procedure for the authorization of capital expenditure (Fig. 4).

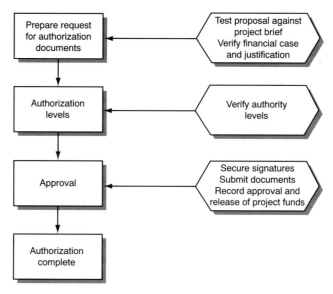

Fig. 4. Authorization process

Request for authorization documents provides the facility for approval of the project proposal in two principal area of authority: (a) within a client's operational/business area structure and (b) within the client company senior managerial/directorate structure.

Authority levels, normally monetary, will be determined and set by the client's capital expenditure procedure. This procedure would need to identify the variance in value of authority levels associated with a projects inclusion or exclusion from the authorized annual operating plan.

Request for authorization documents

Capex request form

Part B of the Capex request form is designed for use as a request for authorization for projects subject to approval at client business area management authority level. Such project proposal costs will be within the prescribed monetary limits and normally within the authorized annual operating plan.

The document will be prepared by the project leader and be based upon data assembled from the project investigation.

Part B of the Capex request form contains sections identifying:

☐ project title and identification
☐ recommendations – brief description of proposed project scope
☐ confirmation of objectives – review of compliance with project brief
☐ justification – achievement of technical and financial objectives

- ☐ project budget cost – estimated capital and expense costs
- ☐ estimated savings – claim of any financial benefits
- ☐ project timescale – planned project milestone dates
- ☐ designated project manager
- ☐ proposal approval signatures – list and signatures of approvals
- ☐ authorization signature – approval signature of client authority holder.

A sample form is given in 'Pre-authorization reporting and documentation', below.

Capex project proposal summary

High value complex projects, subject to approval at senior management or director authority level, call for a more detailed appraisal of the project proposal. Such projects are again subject to the prescribed limits and authority levels and will include any proposals that make significant additions or variations to the annual operating plan.

Request for authorization is by way of a set of standard form documents summarizing the project proposals, identifying responsibilities and considering in greater depth the financial viability of the proposed expenditure. The main document headings are listed as:

- ☐ Description and justification.
- ☐ Project cost analysis.
- ☐ Levels of authority.
- ☐ Responsibilities and risks.
- ☐ Assumptions and analysis of alternatives.
- ☐ Budget analysis.
- ☐ Project timing.
- ☐ Resources.
- ☐ Approvals.
- ☐ Authorizations.

A sample set of forms together with outline definitions are provided in the following section (see Fig. 9).

The project leader is responsible for the preparation, submission and progress of the request for authorization document throughout the authorization process. Securing signatures approving the proposal ensures that all parties within the client's business areas are familiar with and support the project.

Completed and approved request for authorization documents are submitted for final signature at the appropriate authority level to secure project authorization.

The formal recording of project authorization is essential to the initiation of project implementation. Release of funds to the project budget, setting up of project cost control methods and interface with the client's financial systems are all addressed at this point.

Pre-authorization reporting and documentation

Investigation progress reports

The progress of all Capex requests through investigation to authorization is monitored and reported at regular intervals. Information on progress against the outlined project programme and the development of the project cost plan is examined monthly for impact on the annual operating plan. Identified revisions to the AOP will be reviewed at the capital planning group meeting and sanctioned by the client at the appropriate management authority level. All changes to the AOP will be recorded by the issue of a revised annual operating plan (RAOP).

Investigation reports fall into two categories determined by the anticipated value and complexity of the project proposal.

High value/complex projects
Generally defined as investigation of proposals anticipating authorization at client senior management/director authority level and requiring development of a full Capex project proposal summary. The report will contain:

☐ investigation status summary (see Fig. 13)
☐ copy of the project plan (see Fig. 5)
☐ progress overview (see Fig. 14)
☐ summary bar chart (see Fig. 6).

Minor projects
Investigation of proposals where anticipated project value will require authorization at client business area management authority level and will proceed to request for authorization through Part B of the Capex request form. The report will contain:

☐ investigation status summary (Fig. 12)
☐ simple bar chart if warranted by project complexity.

Sample reports for both investigation categories are contained in the following section on documentation.

Documentation (Figs 5–14)

Project plan
This is a graphical representation of the project phases conceived in

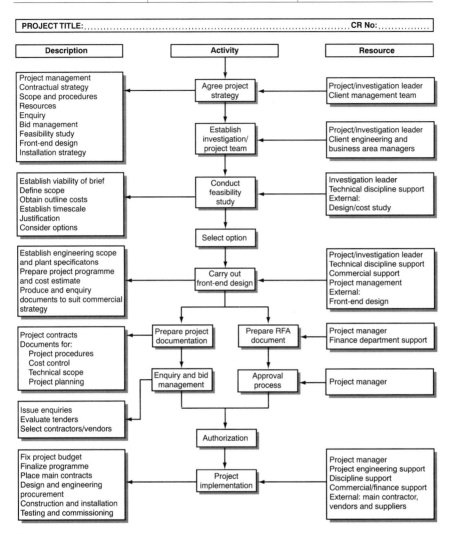

Fig. 5. Project plan

development of the project strategy in flow chart format outlining the plan activities, activity description, responsibilities and anticipated resource requirements.

Outline scope and procedures document

The broad purpose of the document is to define in draft form the project description, technical and process scope, procedures for implementation and project administration. The document will be developed and finalized during investigation and issued at commencement of project implementation.

Capital request form
Part A. Proposal for capital expenditure
Part B. Request for authorization for minor projects.

Capex project proposal summary
Request for authorization of major projects.

Design brief
Full process and technical design requirements presented as deliverables to front-end design process.

Master control estimate
Document to facilitate development of a cost plan during investigation and to monitor expenditure during project implementation.

Investigation progress report formats
High value/complex projects.
Minor projects.

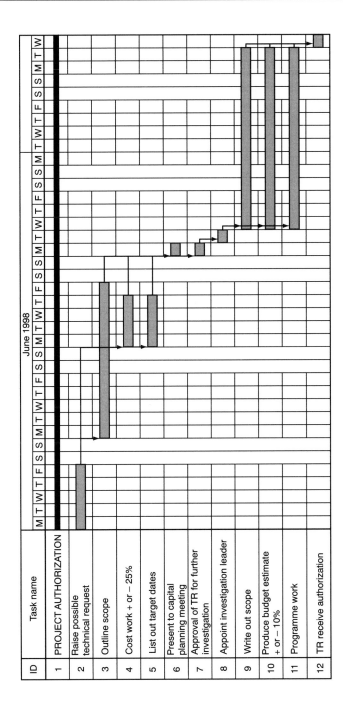

Fig. 6. Pre-authorization programme

SCOPE AND PROCEDURES DOCUMENT

The broad purpose of the scope and procedures document is to define:

- the process/technical scope of the project
- the methods of project implementation
- the procedures associated with implementation
- the roles of personnel required for implementation
- the documentation associated with the project.

As a minimum requirement the following matters will be defined within the document:

Technical/process scope

Project definition and objectives.   Process/production descriptions.
Plant identification.   Services requirements.
Control philosophy.   Building description.
Battery limits conditions.   Enabling works.

Implementation

Project management.   Procurement strategy.
Design responsibility.   Work packaging.
Contractual arrangements.   Project timetable.
Budget control.   Insurance.
Procurement.   Construction.
Safety.   Administration.

Procedures

Project files.   Project planning.
Drawing preparation/approval.   Instructions.
Variations.   Expediting.
Meetings.   Reporting.
Payments.   Cost control.
Quality control.   Testing and commissioning.

Personnel

Management structure.   Client liaison.
Process and engineering discipline   Project team.
resources.   Safety management.

Documentation

Regulations for contractors.   Correspondence.
Information transmittal.   Site instructions.
Scope change.   Variation orders.
Internal reports.   Contractors reports.
Meeting minutes.   Payment certificates.
Take-over certificates.

Fig. 7. Scope and procedures document

CAPEX REQUEST FORM	**CR No.** ....

PART A.

CAPEX REQUEST SUMMARY.

Project title :
Business Area :
Area Manager :
Date :

Scope/Description :

Objectives :

Expected Benefits :

Justification :

Estimated Costs –
        Capital :
        Expense :

Proposed Timescale –
        Investigation Completion :
        Anticipated Authorisation :
        Project Completion :

INVESTIGATION.

CR Allocated to : ....   **Date :** ....

CR Document Issued to : ....   **Date :** ....

CAPEX REQUEST APPROVAL.

Business Area Manager : ....

Signature : ....   **Date :** ....

Fig. 8. Capex request form

PART B.

REQUEST FOR AUTHORISATION.

Project Title:

Recommendation:

Confirmation of Objectives:

Justification:

Budget Costs:
       Capital   £ .... (199..)   £ .... (199..)   £ .... (199..)
       Expense £ .... (199..)   £ .... (199..)   £ .... (199..)

Estimated Savings:

Payback:

Forecast Authorisation (Date) :

Forecast Project Completion:

Designated Project Manager:

REQUEST APPROVAL.

Approval by: ....   **Signature** ....   **Date** ....

             ....   **Signature** ....   **Date** ....

             ....   **Signature** ....   **Date** ....

AUTHORISATION.

Authorisation Observations:

Signature ....   **Date** ....

Fig. 8. Cont.

CAPITAL PROJECT PROPOSAL SUMMARY

FORM. A.

PROJECT TITLE : ....

CR No : ....

Business Area :

Project Leader :   **Date :**

Authority Level :

Proposal Description.

*A brief, yet concise description of what the proposal
will achieve and where*

Justification.

*An explanation defining the reasons and logic
for advancing the proposal*

Fig. 9. Capital project proposal summary

CAPITAL PROJECT PROPOSAL SUMMARY

FORM. B.

PROJECT TITLE : ....

CR No : ....

Project Cost Analysis.

Project costs and returns.
Summary of all costs and financial benefits,
arising as a result of the proposal

Risk and Responsibility.

Identify major risks likely to affect the project.
Environmental issues, technical/process problems.
Industrial relations and statutory obligations

Assumptions and Alternatives.

Identify any assumptions made in the proposal.
State alternative proposals considered and reasons for rejection

Fig. 9. Cont.

CAPITAL PROJECT PROPOSAL SUMMARY

FORM. C.

PROJECT TITLE : ....

CR No : ....

Budget Summary.

Budget costs displayed in tabular form.
Costs identified reflect subheadings and totals from
Project master control estimate

Project Timing.

List of planned dates for principal project activities.
Identify milestone dates for key events and project completion

Resources.

Identify both internal and external resource requirements

Fig. 9. Cont.

CAPITAL PROJECT PROPOSAL SUMMARY

FORM. D.

PROJECT TITLE : ....

CR No : ....

Approvals.

Seen by	Signature	Date
....	....	....
....	....	....
....	....	....

The project manager is responsible for the circulation and completion of the approvals form by all client business area, technical and financial managers having a particular interest in the project proposal.

....	....	....
....	....	....
....	....	....

Project Manager.
....

Authorisation.

Capital Expenditure Amount :

Annual Operating Plan Value :

Authorised by	Signature	Date
....	....	....
....	....	....

Fig. 9. Cont.

CAPEX REQUEST FORM – APPENDIX 1

PROCESS DESIGN BRIEF – DECLARATION

CR No. :

Investigation Leader :   **Date :**

Declaration.

The Feasibility Study Deliverables accompanying this form are accepted as the Project Process Design Brief.

Acceptance is to be viewed as authority to proceed to Front-end Design using the Design Brief as the basis for the Project Scope.

Signatures.

Business Area Manager.   ....   Date...............

Technical Manager.   ....   Date...............

Engineering Manager.   ....   Date...............

Notes.

− Responsibility for obtaining approval and signatures lies with the nominated Investigation Leader
− Preparation of the Feasibility Study Deliverables will be undertaken by the appropriate Business Area/Department and approved by the respective area Manager. Collation of the deliverables into the Project Design Brief is the responsibility of the Investigation Leader.
− On completion of Feasibility Study and approval of the Project Design Brief, the assembled deliverables will pass to the Project Leader nominated for Front-end Design, Investigation and preparation of Request for Authorisation documentation.

Fig. 10. Capex request form – Appendices 1–3

CAPEX REQUEST FORM – APPENDIX 2

FEASIBILITY STUDY – DELIVERABLES SUMMARY

CR No. :

Investigation Leader :   **Date :**

Deliverables.

Description	Prepared by	Approved by
Investigation Timetable	. . . .	. . . .
Process Flow Sheet	. . . .	. . . .
Process Description	. . . .	. . . .
Process Specification	. . . .	. . . .
Outline Plant Specifications	. . . .	. . . .
Process Control Philosophy	. . . .	. . . .
Specific Production Requirements	. . . .	. . . .
Maintenance Requirements	. . . .	. . . .
Energy/Utility Requirements	. . . .	. . . .
Environmental Issues	. . . .	. . . .
Safety Issues	. . . .	. . . .
Commissioning Requirements/Resources	. . . .	. . . .
Training Needs Identified	. . . .	. . . .

Fig. 10. Cont.

CAPEX REQUEST FORM – APPENDIX 3

PROJECT SCOPE APPROVAL

CR No. :

Investigation Leader :   **Date :**

Declaration.

It is accepted that the Project Proposal as submitted reflects, as far as can be reasonably achieved, the Process Design Brief approved as the basis for Front-end Design.

Signatures.

Business Area Manager.   ....   Date...............

Technical Manager.   ....   Date...............

Engineering Manager.   ....   Date...............

Notes.

- This form is to be completed at finalisation of design and subsequent preparation of the Project Proposal. The completed form is to be appended to the Request for Authorisation documentation.
- The submitted Project Proposal accompanying this form contains:
     The Process Scope
     Process Flow Sheet/P&ID
     Plant Specifications
     Process Control Philosophy
     Gen. Arrgt. Drawings (where appropriate)
     Preliminary Project Programme

Fig. 10. Cont.

MASTER CONTROL ESTIMATE

PROJECT TITLE : ....

CR No. ....

Description	Capital			Expense		
	Labour	Materials	TOTAL	Labour	Materials	TOTAL
ON-SITE						
Enabling Works						
Plant & Equipment						
Building & Civil Works						
Mechanical Installation						
Electrical Installation						
Instrumentation & Control						
Commissioning						
Training						
Capital Spares						
Client Stock & Labour						
ON-SITE SUB TOTAL						
OFF-SITE						
Project Management						
Pre-authorisation Costs						
OFF-SITE SUB TOTAL						
CONTINGENCY						
TOTAL TO REQUEST FOR AUTHORISATION		**CAPITAL**			**EXPENSE**	
Latest Forecast		**CAPITAL**			**EXPENSE**	

Fig. 11. Master control estimate

INVESTIGATION STATUS REPORT

MINOR PROJECTS

PROJECT TITLE: ....

CR No. ....   **CUT-OFF DATE:** ....

INVESTIGATION COMPLETION. PLANNED (Date): FORECAST (Date): PROGRESS TO DATE (%):	INVESTIGATION BUDGET. APPROVED/ESTIMATED (£): COMMITTED TO DATE (£):
DESIGN.	ENGINEERING SCOPE.
SPECIFICATION.	CRITICAL ITEMS.
MAJOR PROBLEMS.	NEXT MONTHS OBJECTIVES.
PROGRAMME/MILESTONE DATES. Plan Actual INVESIGATION START: AUTHORISATION: COMMENCE INSTAL'N.: PROJECT COMPLETION:	PROJECT FORECASTS. TOTAL PROJECT COST (£): Capital (£): Expense (£):

....	....	DATE : ....
ORIGINATOR	SIGNATURE	

Fig. 12. Investigation status report – minor projects

INVESTIGATION STATUS REPORT

MAJOR PROJECTS

PROJECT TITLE : ....

CR No. : ....   **DATE** (MONTH/YEAR)

INVESTIGATION STATUS. Overall Progress (%) Investigation Costs (£)	**Planned**	**Actual**

FORECASTS.
 	
Investigation Completion (Date)
Authorisation (Date)
Total Project Cost (£)
    Capital Cost
    Expense
Project Completion (Date)
	CAPEX Plan	**Latest Forecast**

OVERVIEW.

CRITICAL ITEMS/MAJOR PROBLEMS.

MILESTONES ACHIEVED.

NEXT MONTH'S OBJECTIVES.

.... INVESTIGATION LEADER	.... SIGNATURE	DATE : ....

Fig. 13. Investigation status report – major projects

OVERVIEW

*Comment on progress and timing against
each listed heading as and when 'current'
to investigation programme*

1. **Project Strategy.**

2. **Scope Definition.**

3. **Feasibility.**

4. **Approval in Principle.**

5. **Front-end Design.**

6. **Enquires.**

7. **Cost Plan/Budget.**

8. **Request for Authorisation.**

Fig. 14. Overview

3. Commercial management

J. Rollinson

Introduction

Commercial success is a significant factor in the outcome of each and every project. This is measured not only in the ultimate cost benefits derived by the client from the project when operational, but also in the management of costs during the implementation of the project. Many projects have 'failed' economically because, in the final analysis, the client paid more than originally anticipated.

There are two key elements in the successful management of project costs: the determination and control of budgets and expenditure, coupled with an appropriate commercial strategy. Each is also linked to the client's general policies in these areas.

Cost control and estimating is addressed in Chapter 4. Commercial management is the subject of this chapter.

Objectives

Safeguard the company's position

An essential feature of sound commercial management is to ensure that the company's legal and financial position is protected at all times. No company would wish to undertake a financial venture or to be contractually committed where the exposure to risk is high.

Common failings for the inexperienced or poorly-advised company in this area of project management are:

- [ ] inadequately defined scope of works, specifications or prices
- [ ] incomplete, inaccurate or contradictory contract documents
- [ ] inappropriate terms and conditions of contract.

These factors will inevitably lead to payments greater than anticipated or may even combine to create litigation.

It is therefore essential that policies and procedures are in place to minimise the risk. Company 'standard' purchasing arrangements may not be appropriate, so the company should recognize the need for knowledge and experience in the field of commercial project management and, where this is not available in-house, seek the appropriate external support

Specialist vendors and contractors may offer 'package' deals which may have some attraction for the under-resourced company, but independent advice should be sought wherever possible.

Compliance with company policies

When a company is undertaking a capital investment programme involving a number of projects, or even a single major project, it may be recruiting new staff, or adopting new practices and procedures, temporary or permanent. In such circumstances, the company should recognize that the new programme, or management structure associated with it, may take on a 'life' of its own.

The company should ensure that established company policies are maintained at all times and adhered to by the project team(s). This principle may apply in a number of commercial areas such as procurement practices, payment arrangements and accounting procedures.

The company should recognize this requirement at the earliest stage of capital investment planning and place in position the appropriate staff and/ or procedures to integrate the new with the existing. The possibility of departures from existing practices should be identified and resolved. For example, the payment arrangements for a supplier undertaking a one-off major purchase is unlikely to be the same as those for regular, long-term suppliers associated with the normal business.

Procurement co-ordination

It is likely that the company will already be a significant purchaser of goods and services. This process will involve established trading arrangements with a variety of suppliers and contractors. These arrangements should be acknowledged and reviewed in the context of the procurement require-ments for the capital investment programme.

A careful analysis of the goods and services required for the planned projects should be undertaken and matched with the capabilities of existing suppliers and contractors. There are two prime considerations in such an analysis:

☐   To obtain the best commercial advantage.
☐   To identify 'horses for courses'.

For example, it is likely that the projects will incorporate a number of features 'standard' to the company in the form of plant, equipment, building finishes, etc. Where satisfactory purchasing arrangements are already in place for such items, these should be continued, or extended to obtain the benefit of quantity discounts and the like.

Conversely, the risk associated with overburdening existing suppliers or contractors should be recognized, however expectant they may be! Clearly, a small, local company which provides a minor maintenance service is not equipped to undertake a multi-million pound project.

Financial probity

The company should recognize that an infusion of capital into an existing management structure, or a new project management organization, may present the risk of inappropriate financial practices, to the extent of possible corruption or fraud.

This risk should be addressed by the extension of existing company review and audit procedures to the new works and the introduction of appropriate checks and balances.

In the general commercial field, sensitive areas are:

☐ the selection and use of vendors
☐ the agreement and authorization of payment to vendors.

With regard to vendor selection, single source procurement should be avoided wherever possible (unless technically or commercially desirable) and suitable competitive tendering procedures adopted (see later). The extension or amendment of orders and contracts should also be controlled.

All payment procedures should be suitably defined and monitored, with a clear distinction between the ordering, receipt and payment for goods and services. Major projects involving substantial contracts may require an independent party to evaluate and certify payment.

Accounting and audit requirements

The commercial management of projects should recognize existing company financial procedures while being sufficiently adaptable to accommodate the specific needs of the projects.

An accounting procedure for the planning, allocation and monitoring of capital funds may be necessary for an extensive investment programme. Cost control procedures for individual projects (see Chapter 6) should interface with company accounting arrangements for the collection and recording of expenditure.

The particular nature of commercial transactions on projects should also be recognized, including

☐ advance/down payments – requiring third party guarantees
☐ manufacturing/fabrication (off-site) payments – requiring ownership and indemnity arrangements

☐ interim progress (on-site) payments – requiring a valuation and certification procedure.

Company and statutory requirements for audit should also be recognized. It may be considered appropriate for the whole capital investment programme to be the subject of independent financial overview, as well as individual major projects.

Risk and insurance

The company should recognize that project works arising from a capital investment programme may introduce additional risks to the company's business. The presence of construction activities on its property, particularly in an operational environment, brings potential hazards of personal injury, accidental damage, fire and the like.

Such risks are generally insurable, but existing company policies may not extend to construction works. Standard contracting arrangements for construction works may address indemnity and insurance requirements and contractors may carry appropriate policies.

Thus, to ensure that there are no 'gaps' in insurance cover, the company's commercial procedures should include a thorough review of insurance arrangements.

Available resources and skills

The question of resources available to the company is covered in Chapter 1. Just as special skills are required for the design and management of projects, particular skills will be required to manage the commercial elements. The company should assess its available knowledge and experience in this field, taking independent advice, as appropriate.

Familiarity with the design and construction process associated with projects is essential, as is knowledge of the various procurement techniques and related commercial documentation. Experience with suppliers and contractors, particularly where different from those generally used, is also an advantage.

Where such skills are not available 'in-house', they may be obtained on a 'hired-in' or consultancy basis. This may be in conjunction with out-sourced project management or by means of an independent specialist, generally a quantity surveyor. Consultancy appointments are referred to later.

Market factors

As with all types of commerce and industry, market factors will be an influence on the capital investment programme, as well as individual projects. Awareness of these factors is an important element of commercial project management, particularly in the strategic planning stage (10-, 5-, 2- and 1-year plans) and the timing of individual projects.

General economic factors such as inflation and interest rates will be an obvious influence, but key industry factors should also be monitored. The construction industry on the whole is cyclical. When general demand is high, tender prices will rise above general inflation levels, usually accompanied by shortages in skilled labour. Conversely, at low points in the market, the astute client will find tender prices at little more than cost.

The company whose projects involve imported materials or plant should consider the implications of currency fluctuations and forward purchase the appropriate currency. Similarly, the prices of imported materials, particularly metals, which are subject to local market factors, should be monitored and advance positions secured.

Statutory requirements

Most companies will be familiar with statutory requirements relating to general commercial matters (Sale of Goods Act etc). However, projects arising from a capital investment programme may touch upon areas of legislation unfamiliar to the company and appropriate advice should be sought.

The construction industry has its own range of legislation, much of it contractor related, but project works requiring significant demolition or construction works may be subject to the Construction (Design & Management) Regulations. This statutory requirement imposes certain obligations upon a client in respect of health and safety throughout the construction process, commencing at the earliest planning stages. Apart from the specialist knowledge required to implement the regulations, the company should also be aware of the associated cost implications.

Materials and plant from overseas may generate import duties or EC administrative requirements.

Project procedures

The company, when establishing a capital investment or development programme, will develop strategies and procedures for implementation of the related projects (see Chapter 1). Commercial management practices should reflect these strategies and procedures and should be available at the appropriate points in the project development process.

This may include the definition and categorization of projects by type and/or size for authorization, documentation, procurement, vendor selection, payment, etc.

Regularization of commercial procedures will not only facilitate the cost effective implementation of projects but will also ensure consistency of approach between projects. This will be a useful management tool in gauging the performance of projects, individually and collectively. Consistency of approach for a company in its dealings with contractors and suppliers is also an advantage.

Technical achievement

An important factor in the overall management of projects is to strike an appropriate balance between commercial and technical factors. Project budgets should make due allowance for costs that will match technical expectations.

Having established realistic budgets, this policy should be maintained throughout the procurement process, avoiding the temptation to buy cheaply at the expense of specification. Conversely, it is not cost-effective to bypass the competitive tendering process for reasons of programme or expediency. The key influence financially should be value rather than cost.

Contracting strategy

The aforementioned factors, which form the basis for sound commercial management of projects, should be drawn together in the establishment of appropriate contracting strategies for the implementation of the capital investment or development programme.

Contracting strategy

Project implementation

Commercial arrangements are an integral part of project implementation strategy (see Chapter 2). There should be liaison between project and commercial management at all stages in the planning and development of a capital investment programme (Capex) and its related projects.

Appropriate commercial procedures and documentation should be agreed and established to match the company's commercial objectives (see above) and to meet the requirements of individual projects. Such details should be reflected in the project Scope and Procedures document.

The overall objective is to match technical expectations with market prices in order to achieve value for money.

The commercial process

The commercial process associated with project procurement may be similar in many respects with the company's general procurement procedures, particularly if engineering and construction works are procured on a regular basis. The basic principles of identifying requirements, sourcing vendors, obtaining prices and placing orders are fundamental to any commercial purchasing process.

However, it should not be assumed that existing procedures, documentation and personnel may be applied to project works. Similarly, the scope and magnitude of project works may be beyond the capabilities of existing suppliers and contractors. The first stage in developing contracting strategy for capital projects is to address this issue.

The company, through its designated capital programme and project management, should ensure that existing arrangements are reviewed for compatibility with project requirements and that appropriate steps are taken to make good any deficiencies. These may include:

☐ out-sourcing of design, project and commercial management support
☐ market research for potential supplier and contractor identification
☐ preparation of appropriate commercial documentation.

These particular features are addressed in greater detail later in this chapter.

While there are many variants on the final method of vendor appointment, the cornerstone of commercial strategy is the selection of suitable vendors ('horses for courses') and realistic pricing arrangements.

This is best achieved by competitive tendering.

Thus, the basic steps of the commercial process for project procurement may be summarized as follows:

☐ Identify scope and specification of proposed works.
☐ Source and prequalify suppliers and/or contractors.
☐ Prepare suitable tender documentation (technical and commercial).
☐ Obtain prices by competitive tender.
☐ Place orders or contracts.

It should be noted that the commercial process may not cease at this point. For large-scale works there is likely to be a requirement for ongoing commercial activities, often referred to as 'post-contract' support to project management.

Capex administrative requirements

Commercial procedures should be co-ordinated with the overall requirements of the company's Capex plan (see Chapter 2).

Firstly, the appropriate commercial arrangements should be in place to meet the timings of both the collective and the individual capital project programme(s). This is particularly relevant

(a) to the obtaining of market prices for the purpose of project budgets and project authorization (see Chapter 2)

(b) to the timely placing of orders to enable works to proceed (see 'Procurement', Chapter 5).

Secondly, the commercial documentation should reflect the requirements of project management in its dealings with suppliers and contractors in areas such as

☐ programming and reporting
☐ expediting and delivery
☐ cost control
☐ scope change
☐ instruction and notification
☐ performance and warranty
☐ dispute resolution.

Procurement of Capex resources

The company should, in its initial review of the human resources required for the Capex plan, identify any shortfalls in its capabilities in the various fields of project management and take the appropriate measures to supplement them. This may apply in the following specialist areas:

☐ general project management
☐ design
☐ planning
☐ cost engineering/quantity surveying
☐ purchasing
☐ field engineers/technicians.

Shortfalls in these areas may be addressed by individual staff recruitment on a temporary or short-term contract basis. This approach gives a degree of flexibility to the company in terms of matching resources to project timetables and switching resources between projects. It also provides problems for the company in attracting staff of the appropriate calibre and retaining them in periods when the labour market is buoyant.

Alternatively, the company may wish to obtain an external 'service' for these requirements, individually or collectively, by the appointment of consultants. This has particular attractions for larger scale project works. The principal benefits associated with this approach are single point responsibility and a degree of certainty of cost. A consultant would be expected to quote a fixed fee or from a fee-scale for defined services.

However, as the client, the company should ensure that any proposed consultant has the relevant knowledge and experience (usually denoted by some form of professional qualification) and should be aware that such skills do not come cheaply!

There is a wide variety of project-related consultancy services available to a potential client, ranging from international, multi-disciplined consultants through individual design/engineering disciplines to specialist services such as planning and cost control. To assist the company new to such appointments, there are various professional bodies which represent organizations of the relevant skills and who provide advice accordingly.

The consultancy appointments to be made by the client are largely determined by the existing skills within the company, the nature of the project(s) to be undertaken and the preferred approach. Whatever the nature of the appointment, an essential feature should be an appropriate definition of the service to be provided, to include

☐ a client brief for the services required (including scope, deliverables and timetable)
☐ the consultant's proposals (defining how the brief is to be achieved)
☐ the consultant's fees (and payment arrangements)
☐ suitable terms and conditions of engagement (usually based on a professional standard).

An important feature of consultancy engagements for a client is the incorporation of professional indemnity insurance by the consultant as a safeguard against error and negligence.

Design responsibility

An important feature of any project strategy, whatever the scope or size, is the identification and clarification of design responsibility.

Design will feature in the various stages of the development of the project, commencing at the initiation and authorization stage (see Chapter 2) and continuing through the implementation stage (see Chapter 5). At each point, the company should decide and allocate responsibility for the design process, the particular allocation then becoming part of the contracting strategy.

The design function is generally comprised of the following stages:

(a) *Project (design) brief.* A relatively simple and concise statement of the proposed scope and objectives, usually prepared by the department or individual within the company who has 'ownership' of the project. At this stage, it is clearly a client responsibility, although, for larger or more complex projects, external input may be required. (Note: For major projects where the company wishes to adopt a 'turnkey' approach (see below), the design responsibility would pass to the chosen contractor at this point and would thereafter incorporate all the subsequent stages.)

(b) *Outline ('front-end') design.* This is the process whereby the design brief is converted into a conceptual scheme incorporating the required buildings, plant, equipment and services. Dependant upon company resources, this stage of development may be undertaken 'in-house' or by external consultants. Alternatively, the company may wish to retain key elements of the design particular to or confidential within its business and to 'contract-out' the peripheral design. For larger scale projects, this stage may take the form of a design study which serves the purpose of establishing the technical viabilty of the proposed scheme. The use of external consultants at this stage will require the appropriate form of appointment, as previously described.

(c) *Detail (engineering) design.* This stage is comprised of the development of working details from the conceptual scheme to enable manufacture, fabrication and construction to proceed. Although this activity may be undertaken 'in-house' or by external consultants, it is customary for the responsibilty to be passed to the suppliers and contractors for the relevant parts. However, any residual responsibility for co-ordination should be noted and addressed.

Allocation of project works

Having identified responsibilty for design, the next key element of contracting strategy is the commercial allocation of the project works. It is essential at the earliest stages in the development of the overall project strategy to reach agreement, technically and commercially, as to the means of implementing the full scope of the project. This is not only essential in the allocation of physical works but also in the determination of areas of responsibility. The agreement of this strategy will also indentify the required workload for design, procurement and administrative activities.

There are four basic options for implementing project works:

(1) *In-house.* The company may have its own labour and/or material resources to undertake the works with minimal procurement of external supplies of goods and services.This approach may be appropriate for small scale works of no great complexity, but it is becoming increasingly rare for companies to retain the capacity for such a commitment.

(2) *Turnkey.* The company may wish for reasons of scale and/or expediency to involve external support to the maximum extent at the earliest possible stage. This may be achieved on a 'turnkey' contract, whereby a contractor is appointed on the basis of a project brief defining general scope and overall performance criteria and undertakes all of the design and construction of the entire facility. A fully operational plant is passed to the client on completion. There is

considerable cost associated with this strategy and it presumes minimum input from the client on matters of detail. For these reasons, this approach is more likely to be found in the petrochemical and other heavy industries.

(3) *Design and build.* This strategy is a progression of the 'turnkey' approach in that it involves the appointment of a single contractor to undertake the design and construction of the project. However, the appointment is preceded by a more detailed consideration by the client of design requirements ('front end') and a more detailed tendering process. Having appointed the chosen contractor and desired scheme, the client has the advantage of single-point responsibility, but limitations in specific detailing and specialist vendor selection.

(4) *Work packages.* Under this approach the client breaks down the project works into areas of specialism, usually on a trade basis (mechanical, electrical, etc.) to suit the nature of the particular project. The client then issues tenders, selects vendors and places orders for each individual work 'package'. This method requires greater client/ consultant input at the tender enquiry stage to assemble the designs, specifications, etc., for the various packages but gives more choice and flexibility (and possible cost savings) in vendor selection. There is, however, an additional responsibility for the client's project manager for co-ordination of the design and construction of the packages. There are variants on this method whereby the client may introduce an overall 'management contractor' to assume some or all of the client responsi- bilities defined here.

Notwithstanding the allocation strategy selected, there may be within the scope of the proposed project works certain key items of supply and/or installation. These may be specialist items of plant or equipment fundamental to the client's process or production or specialist contractors employed on a regular basis. The client should retain these works on a 'direct supply' basis, making provision for integration within the other contracted project works.

Project completion procedures

When considering the contracting strategy for project works, there is a tendency for clients to focus on the earlier stages of the project development, namely design and procurement, leading to the construction phase.

Of equal importance, particularly to a process or production company, is the manner in which the plant is brought into operation. This phase may include a variety of activities such as testing, commissioning, acceptance/ take-over procedures and performance trials. It may also extend to warranty arrangements.

Each of these activities requires careful definition, technically and commercially, in order to properly assign obligations and responsibilities. The strategy to be adopted by the client may be dependent upon the nature and complexity of the project works, the extent of technical skills required and available and the timetable for operation.

Vendor selection procedures

At the earliest stage in the development of contracting strategy, the company should decide upon the basic selection process for suppliers and contractors. This aspect of strategy may, to some extent, be determined by existing company policy on such matters.

It is common commercial practice to procure goods and services by means of competitive tendering. This practice holds good, and may even be considered essential, for capital expenditure programmes, particularly if the works are of significant size. This approach will generally give a client the comfort of securing the best 'market price' for the particular works.

However, there is a view that competitive tendering is a lottery, particularly when the project works are poorly defined or the bidding vendors are of indeterminate status or capability. Also, when the tendered works are complex, there is a risk that the successful tenderer will be the one making the most pricing errors. This may lead to problems of quality, time or cost when the works are implemented.

Aside from risks arising from the vendor appointment, the company should also be aware of the resources and time required to effect competitive tendering. To reduce this problem, the company should determine a minimum level of implementation of, say, £5000, although there should be a degree of flexibility in terms of the nature and complexity of the proposed works.

If the company is a regular purchaser of goods and services which are akin to the proposed project works, and has an established and reliable vendor base, a strategy of negotiated prices may be desirable. However, this method is only cost-effective when the client has a reliable means of gauging prices and assessing 'value for money'.

Another method of negotiated tendering is referred to as 'partnering'. There are varying interpretations of this procedure, but in general terms, it relates to major companies with large capital expenditure programmes who are able to engage a small number of vendors throughout the programme on a pre-selected or negotiated basis. Apart from economies of scale, such companies may derive benefit from closer working arrangements at the earlier stages of projects, jointly developing more cost-effective schemes.

The competitive and negotiated methods of procurement may be combined in a two-stage tendering procedure. Under this procedure, competitive tenders are obtained on a preliminary or outline basis at the first stage. The prices are thus indicative or budgetary only and may be accompanied by technical proposals. The successful tenderer is then invited to enter into negotiations on the final price for the detailed works at the second stage. It should be noted that in these circumstances the successful first stage tenderer is at an advantage. This risk can be reduced by establishing a 'not to be exceeded' tender sum arrangement.

The process of tendering is described below in 'The tender process'.

Vendor status

If within the overall commercial strategy the company determines that certain suppliers or contractors are to be engaged on key works, the status of these vendors should be clearly defined. The following definitions may apply:

☐ *Directly employed.* Engaged upon specialist works within the overall scope of a project being undertaken by another (main or principal) contractor. Directly employed by the client rather than being a subcontractor to the main contractor.

☐ *Nominated.* Provision is made within the scope of works by a main contractor for certain elements to be undertaken by a specialist selected (nominated) by the client. The provision is made by means of a pre-stated allowance within the main contract. The nomination is made by the client subsequent to the appointment of the main contractor. The specialist, when selected, becomes a subcontractor or supplier to the main contractor. While the nomination is made solely by the client, the nomination process is defined and agreed with the main contractor.

☐ *Specified.* Similar to nomination, the difference being that the specialist is selected prior to the appointment of the main contractor and is identified within the main contract documentation. The specialist becomes a subcontractor or supplier to the main contractor.

☐ *Approved/preferred.* Specialists identified to the main contractor by the client as being suitable or desirable for particular works without the inflexibility and formality of nomination or specification. Selection is at the discretion of the main contractor and the specialist becomes a subcontractor or supplier to the main contractor.

Commercial documentation

Integral with contracting strategy is the selection of appropriate commercial documentation. The company should determine whether existing company forms, particularly terms and conditions of contract, are adequate for the purpose of the proposed project works.

It is recommended that reference is made to the various 'model' forms of contract which prevail in many industries (mechanical, civil engineering, etc.). These have the advantage of familiarity to the suppliers and contractors within those industries. They also serve a useful purpose in establishing an overall framework of contract documentation, integrating specifications and the like. The client, of course, is at liberty to modify these forms for his particular purposes.

Confidentiality

The company should be aware that during the development of a capital investment programme, commercially sensitive business issues may be identifiable. The company would wish to protect the confidentiality of such matters.

It is therefore advisable to introduce provisions for confidentiality into commercial strategy at all stages of a project development where such issues arise. This would commence with consultancy agreements, continue in tender enquiry documents and be included in the relevant contracts and orders.

The provisions for confidentiality should include arrangements for the exchange and retention of information, such as process and production data, plant drawings, etc. There may also be a requirement for indemnity for breach of confidentiality from the recipient of such information.

Copyright is another issue likely to arise in the development of designs and selection of equipment for a project. The commercial strategy should ensure that the rights of the respective parties are clearly defined. This can be a complex matter and the appropriate legal advice should be obtained. It should be noted that these matters are also addressed in the 'model' forms of contract employed in many industries.

Forms of contract

Contract documentation

Closely linked to contracting strategy for projects is the selection of the appropriate contractual documentation. The basic objective here is to achieve proper definition of the responsibilities, rights and duties of the contracting parties.

It is essential that such documents should both encapsulate and safe-guard the technical and commercial expectations of the company. There should also be a balance between the technical and contractual elements – a 'watertight' set of contract terms and conditions may be ineffective if the technical content is poorly defined. The matching of these elements should include the interfaces of technical and commercial factors, such as performance obligations, payment arrangements, etc.

The approach to documentation should be sufficiently flexible to reflect the technical scope and financial value of the particular project works. While key commercial factors, such as quality, price and time should be addressed on a consistent basis, the volume of documentation should match the scale and complexity of the project objectives.

When developing the contracting strategy for project implementation, the company should review its existing commercial documentation, including contract terms and conditions, for suitability for the proposed project works. If necessary, guidance should be sought from legal advisers or consultants within the relevant professions.

Work packages

A key factor within the determination of project contracting strategy is the allocation of the project works, including design responsibility (see 'Contracting strategy'). Contract documentation should be linked to the selected 'packages' of work.

If single point contracting responsibility is desired, including design, this strategy may be developed commercially as a 'turnkey' or 'design and build' contract (as defined in 'Contracting strategy'). These contracts are multi-disciplined, encompassing all the technical skills required for the project. Generally, the emphasis in such documents is on the project as a whole, particularly the performance of the end product. Less attention is thus required on details (specification), although the client should always ensure that his basic requirements are adequately defined.

Where the client wishes to place greater emphasis on specification and has the resources for closer control of detail, the project may be comprised of a range of single-discipline contracts. These documents may be assembled individually to suit the relevant trade package, but may have common core elements.

Time and cost restraints

While technical performance, quality and other end-user requirements may be key elements in the project outcome, the company may also be obliged to consider the other significant factors of time and cost.

A sound principle of any Capex programme is the objective for projects to be 'fixed' in terms of scope and price. Contract documents are then drafted accordingly. However, for this process to be effective, it requires attention to detail and thus time and resources. Where these are not available, exposure to risk of cost escalation is increased. The company may recognize this and adopt a contracting strategy accordingly.

The risk of potential cost increase may be minimized by developing as much detailed design as possible in the time available. Minor changes of detail may then be addressed as the works proceed. Thus, as the scope of the works is not fixed, the quoted price may be subject to adjustment. Apart from modifications to detail, additional costs may also arise from movement in market prices of labour and materials. These factors are addressed in the drafting of the contract documents.

In circumstances where the company is only able to produce minimum details and time is a critical factor, the fixed price strategy may be abandoned in favour of a 'reimbursable' approach. Under this strategy the full design is developed as the work proceeds and the selected contractor is reimbursed the full cost of the labour, materials and plant employed. The contract documents are drafted accordingly. It should be noted that this form of contracting carries the highest risk of cost uncertainty.

Model forms of contract

While the company's general terms and conditions of trading may be suitable for smaller project works, it is likely that a more comprehensive commercial document may be required for larger scale projects.

The company is, of course, at liberty to produce its own contract document by expanding existing trading conditions, acting on legal advice. However, 'home-made' contract terms and conditions tend to be one-sided and may present difficulties over acceptance and interpretation by vendors.

A more common approach is to employ the standard or 'model' forms of contract used in various industries. These documents are generally published by professional bodies representative of the particular industry. They are issued in the form of a set of standard terms and conditions for use on project works in that industry and are often supplemented by related schedules, appendices, options, etc. The advantage of the standard forms is that they are drafted by committees comprised of the various diverse organizations operating in that particularly industry, principally clients, contractors, suppliers and the professions (engineers, architects ,etc.).

The standard or model form thus becomes a balanced document representing the consensus of the views of the various contributing bodies and is widely accepted for use on project works within that particular industry. However, it should be noted that, like any document formulated by a committee of differing views, there may be elements of compromise in certain areas upon which the company may have its own views. In fact, clients often adapt model forms to suit their own particular needs, but care should be taken to avoid inconsistencies caused by modification. Extensive change is not recommended as it may destroy the balance of the document.

A further advantage of model forms of contract is that they address most if not all of the issues that are likely to occur in the out-workings of a project (quality, time, price, performance, etc.). They also provide a legal framework for associated project documentation such as specification, tender, schedule of prices, procedures

There is a wide variety of model forms of contract available for project works and the following sections identify those commonly used in various industries in the UK and often overseas. A client involved with projects or vendors overseas may also encounter various standard sets of terms and conditions issued by a variety of national or international trade bodies. Such documents should be viewed in the context of equivalent UK documents and British Law, with the appropriate legal advice. Further advice on the particular model forms described hereinafter may be obtained from the relevant professional body:

Process industries – documents published by the Institution of Chemical Engineers:

☐ The 'Red' Book – suitable for large-scale turnkey projects (see previous description) for complete process plants (usually petrochemical) on a fixed price basis.

☐ The 'Green' Book – suitable for turnkey projects (as above) on a cost reimbursable basis (see previous description).

Mechanical and electrical industries – documents published by the Institution of Mechanical Engineers and the Institution of Electrical Engineers:

☐ Model Form MF/1 – suitable for the design, supply and installation of new plant or plant replacement in a variety of applications.

☐ Model Form MF/2 – suitable for the supply only of major items of plant or equipment with an option for supervision of erection.

☐ Model Form MF/3 – suitable for the supply only of general items of plant or equipment.

Civil engineering industry – documents published by the Institution of Civil Engineers:

☐ New Engineering Form – suitable for a wide variety of project applications combining civil/mechanical/electrical construction works – based on a 'core' set of conditions with a range of optional documents to suit the commercial approach to the particular project.

☐ Standard forms – suitable for large scale civil engineering works.

☐ Minor Works – suitable for small scale engineering works.

Building industry – documents prepared by the Joint Contracts Tribunal and published by the Royal Institute of British Architects:

☐ JCT Standard Form – suitable for large scale building works.

☐ JCT Intermediate Form – suitable for medium scale building works.

☐ JCT Minor Works – suitable for small scale building works.

Information technology and telecommunications – documents published by the Chartered Institute of Purchasing & Supply:

☐ Model P – suitable for supply and installation of computer equipment, and available with variants for software development.
☐ Model T – suitable for supply and installation of telecommunications equipment.

Consultancy appointments

Where the company wishes to create or to supplement design and project management resources by the engagement of consultants, this may also be undertaken on a contractual basis.

There is a variety of standard forms of appointment for consultancy services available from various professional institutions, including the following:

☐ Association of Consulting Engineers – suitable for the engagement of engineers of various disciplines (mechanical, electrical, structural, etc.).
☐ Royal Institute of British Architects – suitable for the engagement of architectural services.
☐ Royal Institution of Chartered Surveyors – suitable for the engagement of quantity surveying services.

These standard forms of appointment are generally structured to allow engagement on a progressive basis, for example outline design, detailed design, site supervision, etc. This allows the client to match the appointment with his contracting strategy and the sequence/programme for the project.

Apart from the terms and conditions of the standard form, other essential elements of a consultancy appointment are the client's brief and the consultant's proposals.

The tender process

The function of tendering

The prime function of the tendering process is to enable the client to secure the best market price for the goods and services required by means of competitive bids on a common basis.This is achieved by the client formally inviting bids from interested parties, clearly setting out the technical and commercial parameters of the tenders required.

It is desirable that the process is undertaken on a formal basis to ensure equality of opportunity for the tenderers and to simplify the appraisal of tenders by the client.

An additional function of the tender process is to provide cost yardsticks at the appropriate point within the overall project timetable (see Chapters 2 and 5) and as a basis for project budgetary and cost control (see Chapter 6).

Vendor sourcing

The selection of vendors is a key factor in the successful outcome of a project and the company's strategy should be determined at the earliest stage. Where the company is unable to draw upon an existing pool of known vendors or is venturing into new technical areas for a particular project, new vendors will have to be sourced. This may be achieved in a number of ways.

If the project is of significant size, the company may advertise for prospective tenderers in relevant trade magazines, providing essential details of the projects and prequalification procedures. This method has the drawback of the company being potentially overwhelmed by speculative and inappropriate responses.

If the company has engaged professional advisers or consultants for design or other aspects of the project, these parties are likely to be able to recommend suitable tenderers, based upon their knowledge and experience of vendors.

There are several trade directories available for various industries whereby prospective clients may source potential vendors. These directories are normally classified by the specialist skills of the vendors.

Many companies are approached on a regular and unsolicited basis by previously unknown vendors with general expressions of interest in providing goods or services, particularly when general development plans are publicized. The company may turn these approaches to advantage by operating a sifting process, identifying, classifying and registering prospective vendors for future reference.

Prequalification of vendors

Having identified potential tenderers, the company must be satisfied that those vendors have the ability to undertake the project works. Also the list of tenderers must be contained to a manageable number, usually 3–4 for a major, complex project, up to a maximum of 6–7 for lesser works. To meet these objectives, a shortlisting process and pre-qualification proceedure may be required.

Prequalification may be undertaken on a questionnaire basis, backed up by personal inspection and interview. Submissions from the prospective vendors may be requested. Questions should cover the general standing of the vendor and background to the particular project works.

General questions should include the structure, resources and financial status of the company in question, as well as their track record on manufacturing/fabrication/construction/safety, etc. Specific questions should relate to previous (and relevant) project experience, with appropriate references from other clients.

Tender enquiry documents

Having determined that a formal tender process is required, this should be supported by a properly assembled tender enquiry document which should clearly define the technical and commercial aspects of the project works. The accuracy and competitiveness of the tenders will be directly related to the clarity and comprehensiveness of the enquiry document.

The scope and volume of the enquiry document will be dependent upon the scale and complexity of the project works. However, it is suggested that the following elements are provided in enquiries for any significant tenders:

(a) *Invitation to tender.* This formally invites the offer and defines its details, including tender submission date, tender validity period and supporting documentation to be provided. It may also cover administrative information such as acknowledgement of receipt and confirmation of bid, queries on details and arrangements for site visits. Matters such as confidentiality may also be addressed.

(b) *Form of tender.* This document is to be completed and signed by the tenderer as the formal offer to undertake the works described for the quoted price in accordance with the prescribed terms and conditions. Unauthorized amendment to this document should not be permitted.

(c) *Conditions of contract.* This sets out the terms and conditions of contract that the client proposes to employ. Where applicable, reference should be made to the relevant model form, stating any options, appendices and amendments.

(d) *Drawings.* The range of drawings to be included will be dependent upon the quantity of design work to be produced by the client and contractor respectively, varying from fully detailed drawings to basic layouts. The number of drawings will be dependent upon the scale and complexity of the works.

(e) *Specification.* This defines the technical details of the client's requirements and should be as comprehensive as the contracting strategy permits. It should include descriptions of the scope of works, relevant standards of materials and workmanship, and design and performance criteria. Reference should also be made to any programming requirements and site operational and safety obligations.

(f) *Schedules.* This covers a variety of documents which generally provide supplementary or supporting details to technical elements, for example equipment, or commercial elements, for example prices.

Tender appraisal

To maintain the principles of equality of opportunity for tenderers and to ensure a comprehensive review of tenders received, the client should approach tender appraisal in a structured and co-ordinated manner.

The approach will vary in detail, dependent upon the size and complexity of the tenders, but key principles should be retained.

On the date set for tender returns the client should assemble technical and commercial representatives from the project team to open and initially record the tender results. To avoid breaches of confidentiality, no tender should be opened until all are received. Similarly, faxed tenders should be avoided. Basic details of the tenders received should be recorded at this point, that is, price, delivery and technical and commercial conformity. No conclusions should be formed nor results published until each tender has been more thoroughly reviewed.

For large-scale project works it may be necessary for the client to assemble a project team to review and evaluate tenders, particularly where varying technical solutions are proposed by different tenderers. To assist in this process, it may be necessary to arrange post-tender meetings with tenderers to review queries, errors and qualifications.

When one, or possibly two, tenderers have been short-listed, it may be appropriate to conduct a final round of negotiations in order to achieve the best possible price. Then the successful tenderer may be notified and, dependent upon the stage reached in the overall project time-table, the contract award made. It is also courteous to advise unsuccessful tenderers accordingly.

4. Cost control and estimating

J. Rollinson

Introduction

Cost control is a corner-stone of successful project management. Together with the control of quality and time, it is an essential ingredient of each and every project. This principle extends beyond the individual project and also applies to the capital investment programme as a whole.

The control of costs is undertaken in two basic steps; the determination of budgets by the use of estimating techniques, and the monitoring of expenditure relative to those budgets. The development and accuracy of the first step is as important as the attention paid and actions taken on the second step.

These steps are addressed in this chapter by description of their objectives, functionality and operation.

Chapter 3 describes how cost control is linked to commercial strategy.

Objectives

Safeguard the company's position

When a company undertakes a capital investment or development programme, it is likely to represent a significant financial outlay for the company, with the possibility of external borrowing requirements or other funding arrangements. The company's senior management (and its financial backers) would therefore wish to see measures in place which will safeguard the committed funds and to provide the desired return on the investment.

A poorly managed capital expenditure (Capex) programme or significant overspend on a particular project may not only cause cash flow problems but may also restrict future profitability and business plans. These factors may also combine to adversely affect the company's position in the market as a whole. Company balance sheets and annual reports are occasionally obliged to make provision for projects that have gone wrong, which brings the problem into the public eye. The company is thus embarrassed internally and externally.

It is therefore essential that cost control procedures are in place which meet the requirements of best accounting practice and provide a management tool for those responsible for the Capex plan and its individual projects.

The two key elements of project cost control procedures are the determination of budgets and the monitoring of expenditure against those budgets. However, the establishment of these elements is not in itself the safeguard for the company's finances. The procedures must also include an appropriate and pro-active reporting, review and follow-up system at appropriate levels within the company. An external audit arrangement may also be considered.

Establish overall plan

Cost control should be commence at the earliest stage of the Capex plan and should be recognized at the highest level of the company. A Capex plan may founder upon ill-conceived budgets or an over-expectation of what may be achieved for the outlay, leading to an inevitable overspend or curtailment of activities before completion.

It is therefore essential that a realistic overall Capex budget is determined prior to commitments being made by the company. The Capex programme is likely to have its origins in the long-term business plan of the company, having strategic, production or profitability objectives. The business plan may also define the available funds. Those responsible for the Capex plan are thus obliged to work within this framework. However, the next stage is a vital one and requires astute judgement. It consists of linking available funds to technical/commercial expectation, thereby creating the budgets.

A key factor in formulating Capex budgets, whether individually or collectively, is the methodology and accuracy of cost estimates. Obviously, the more effort that is employed on this process, the greater the level of accuracy. This is, of course, a time consuming activity and the company should balance the requirements and formulate a policy on permitted levels of accuracy. Such a policy may be progressive, that is related to the various stages of the Capex authorization procedure (see Chapters 1 and 2). For example, the permitted levels of accuracy of estimate may be as much as $\pm 50\%$ for the initial overall plan and down to $\pm 10\%$ prior to commencement.

The judgement on accuracy of estimate for the overall Capex plan is also related to the scope, complexity and technical risk of the proposed projects, as well as external factors, such as inflation. Whatever the final judgement on accuracy, the senior management of the company should be aware and plan accordingly

The overall Capex budget is thus created by the sum total of the cost estimates for the individual projects which comprise the Capex plan. This budget may then be time related by prioritizing the projects and linking to funding availability.

Co-ordination of financial procedures

Cost control for a capital investment programme will require procedures particular to the nature of project execution, such as the duration of activities, payment arrangements, etc. These may differ somewhat from the company's normal revenue related accounting activities and general management accounting systems.

However, an objective of Capex cost control procedures should be co-ordination and integration with prevailing company accounting procedures and systems. This should be a two-way function such that expenditure on capital projects is properly accounted for and that appropriate cost data is fed back to project management. The interface between these functionalities should be reviewed at the earliest stage.

For example, having established budgets to operate the Capex plan, the company's accounting system should establish cost centres to capture related expenditure for the purpose of cost monitoring. Co-ordination is essential as the company would not wish to go to the trouble and expense of setting-up parallel accounting systems. Neither would the project manager wish to trawl through a mass of unrelated data to identify and retrieve cost information.

Project accounting is addressed in greater detail in Chapter 6.

Information technology

It is likely that the company's accounting systems will be computer-based. The facility for project cost control should be investigated (see Chapter 6).

There is also a variety of software aids for commercial project management on the market. These range from cost estimating databases through to cost management packages. The suitability and compatibility of such systems should also be examined.

Available resources and skills

The question of resources available to the company is covered in Chapter 1. As project cost control is a particular skill, requiring familiarity with the construction process, the company should assess its resources in this field and plan accordingly. Independent advice may be obtained from a quantity surveyor.

If the proposed projects are large and complex, with a high degree of site installation, expertise may be required to formulate the cost estimates. This function may be linked to the development of design and subsequent procurement process (see Chapter 3).

The requirement for this expertise is likely to continue into the project execution phase on matters of cost evaluation and cost reporting (see

Chapter 5). On major process plant projects, this function is often undertaken by specialist cost engineers. In the general construction industry, quantity surveyors are often employed

Development of project budgets

Having determined the overall Capex budget (see previous chapters), the company should then address the detailed budgets for the individual projects. The detailed project budget may fulfil two functions:

(a) to establish the economic viability of the project and thus to form part of the authorization process (see Chapter 2)
(b) to provide a yardstick for cost control as the work proceeds.

The process of establishment of the project budget may have already began with the overall Capex plan (see Chapter 2). An initial cost estimate may have been identified with a predetermined level of accuracy. The next stage is linked to the development of the project design which will provide further information for the refinement and greater accuracy of the cost estimate. At this stage, the estimate is regarded as the cost plan.

When there is sufficient confidence in the accuracy of the cost plan, the company may then review the financial objectives of the project for the purposes of justification for final authorization. This process would generally be an investigation of affordability and return on capital investment. However, in some circumstances, there may be other justifications, such as time-expired plant, health and safety or environmental issues or new statutory requirements. During this process cost is, in effect, converted to value.

The assembly of the cost plan should be undertaken in such a way as to provide a basis for cost monitoring once the project has been authorized. This may be achieved by allocating the various elements of the cost plan to predetermined cost centres which will match the cost expenditure captured by the company's accounting systems (see Chapter 7). At this stage, the cost plan becomes the fixed project budget.

Project cost management

The responsibilities of the project manager on cost matters at the pre- and post-authorization stages of the project are identified in Chapters 2 and 5 respectively.

An overall objective for the project manager should be the establishment of integrated cost control procedures whereby actual cost commitment and expenditure are related to the fixed project budget. Resources and systems should be in place to support these procedures.

During the assembly of the project cost plan, the project manager should ensure that all elements of the project are adequately provided for in the estimates, particularly the interfaces. Compatability with technical and performance requirements should also be confirmed.

Following project authorization, the project manager should review expenditure relative to the project budget on a regular basis. Reporting procedures should be in place, the project manager being the link between the detailed reviews required at site level and the overviews required by senior management of the company. On major projects, contractors' cost reporting procedures should be co-ordinated to suit the client's requirements.

Functionality

Financial operating plans

The requirements for an overall cost plan for the proposed capital investment programme are described in Chapter 6. For major, long-term programmes, a structured series of plans may be developed, ranging from a 10-, 5- or 2-year plan, down to an annual plan. The function of these plans is to assist in the company's overall financial planning. The longer the duration of the plan the greater the benefits to the company's economics.

The funds allocated to the overall plan may be the sum total of all of the project cost estimates or a fixed contribution from the company's general business economic plan or an amalgamation of the two.

Plans of 10 or 5 years duration are regarded as long-term strategic plans and thus subject to revision to suit changes in the company's general trading position, as well as external influences such as economic factors.

The 2-year plan should be regarded as an operational plan, bearing in mind that projects of significant size are likely to be of at least 2 years duration from inception to completion.

The annual operating plan is of the greatest significance to the company, as it will determine exact requirements for cash flow, resources, etc. It is also significant for the project manager, as it is likely to determine which projects will be activated, although it is possible that much expenditure within a given year may, in fact, be carry-forward of commitments on projects commenced in the previous year.

Identification and allocation of projects

Having established the overall Capex plan, the desired projects should be appropriately allocated within the timetable. This process requires two elements of financial planning, that is, an estimate of likely project cost and a prioritization procedure.

An initial estimate of project cost is required in order to determine its appropriate slot within the Capex plan. For example, four similar projects each of £3 million could not be accommodated within an annual plan set at £10 million. Similarly, the company would not wish to commit costs, time and resources to pursuing an ill-defined project, the costs of which could vary by 100%. The principles of accuracy of the initial cost estimate are addressed in Chapter 6. The out-workings of cost estimates are described later.

Prioritization is a process basically linked to the business needs of the company. The needs may be determined by a variety of sources within the company. Sales and marketing may require new product lines. Production may wish to replace worn plant or to increase production. Health and safety or environmental issues may require new methods. The assessrnent of priority may also be linked to the relative justification for each project. Those projects with limited or marginal benefits will obviously be displaced by those providing a good return (see also below). However, a requirement to comply with new legistation may take precedence. It is therefore desirable for the company to establish a review committee to assess competing bids from different parts of the business and to establish priorities.

Authorization procedures

Notwithstanding the allocation of a project to the Capex plan, financial justification may be still be required. The initial inclusion of the project in the plan may be based on an identified business need. However, this need should still be justified and this may be confirmed by a formal project authorization procedure (refer to Chapter 2). The authorization procedure may also be linked with the prioritization process (see above) which should provide consistency of approach within the company's management structure.

The development of the project scheme and its associated cost estimates will provide a base case for justification. The capital outlay thus identified may be compared with the anticipated financial returns for the project. At this stage, the cost estimate would be at advanced level and is referred to as the project cost plan. The preparation of the cost plan is described in greater detail later.

However, before the authorization stage is reached, a preliminary estimating function is required to establish basic economic viability. This is described below.

Preliminary cost plan

When a project is instigated, the scope may lack definition and the data may not be available to support the business case. It is therefore essential at this stage to prepare a preliminary estimate to establish the basic economic viability of the project. This process may also be formalized within the overall Capex operating arrangements as preliminary approval of the scheme, or approval in principle.

The preliminary cost plan (sometimes referred to as 'first order of costs') may be assembled in a number of ways. This will be dependent upon the technical nature and scope of the project and the amount of design information available. This procedure is also known as approximate estimating. Approximate estimates should be suitably weighted to allow for any incomplete information and/or risk factors.

The preliminary costs established by approximate estimates may be reviewed in terms of the company's overall financial expectations of the project, that is, affordability, and thus may the viability of the project be confirmed.

Project cost plan

Having established the overall viability of the project by way of the preliminary estimates, the full project cost plan may be developed. The cost plan fulfils two functions – financial basis for project authorization, and budget for cost monitoring during project execution. Attention should thus be paid to the detail of this cost estimate, not only in terms of level of accuracy, but also in structure and assembly.

The level of accuracy of the cost plan is directly related to the amount of detail (and time) available for its preparation. It is a sound principle of cost control for capital expenditure that the scope of the project is frozen at this stage (and signed-off, where appropriate) and, sufficient design work undertaken to enable detailed estimates to be prepared. The preferred method of obtaining estimates is by way of fixed price quotations from contractors and suppliers.

The structure of the estimate is generally determined by the proposed project contracting strategy (see 'Contracting strategy' in Chapter 3) and the company's desired cost monitoring and reporting format. A typical structure may be comprised of:

☐ *On-site costs.* Plant, materials and labour required for the manufacture, fabrication and installation of the project works – obtained from contractors and suppliers, with prices determined through fixed price quotations – may be broken down on an elemental or trade basis (mechanical, electrical, etc.).

☐ *Off-site cost.* Design and project management services (including consultancies) – prices obtained by way of internal estimates and fee quotations.

☐ *Contingency.* Allowance for risk factors.

The assembly of these costs is described in greater detail later. The completed cost plan is then incorporated into the project authorization process (see 'Authorization process', in Chapter 2).

Establishment of project budget

Upon authorization of the project, the cost plan is converted into the project budget for cost monitoring and reporting functions. If the cost plan has been structured and assembled in an appropriate manner. the transition should be relatively simple.

The initial task of the project manager and/or cost engineer would be to review the various elements of the cost plan for compatibility with the proposed project execution plan, making appropriate adjustments where necessary, for example changes in the 'packaging' of contracted works. Such changes should, of course, be regarded as inter-budget transfers and must not depart from the authorized project sum.

Also, the company's cost accounting procedures should be reviewed to ensure that costs incurred are properly matched to the project budget. This may require the establishment of special cost centres.

The two elements of cost control, budget and expenditure, are drawn together in the cost reporting function (see below).

Cost reports

The management tool generally employed for cost control of projects is the cost report. This type of report has a dual function – as a means of capturing and recording the project costs for the project manager, and to communicate that data, in full or in summary form, to the company's senior management.

The style and content of the cost report will be dependent on the specific needs of the project manager to monitor and control his costs and the particular approach of the company to feedback on the financial status of projects in general. A typical report will be produced on a monthly basis, structured around the build-up of the project budget (cost centres) and comparing costs incurred relative to those centres. The result of this comparison will be a series of individual and collective cost variances which, if monitored on a regular basis, will provide a detectable trend.

This basic approach to cost monitoring is historical in nature. It is equally important that the format of cost against budget is repeated as a forecast of future trends, culminating in an estimated final out-turn cost.

The structure of the cost report should also be flexible to reflect changes in the scope and timing of the project as the work proceeds and to monitor the impact of these changes on the out-turn costs. Change management procedures are covered later.

Another feature of this style of cost reporting is the facility to provide a cost monitor of the design and procurement functions, as well as the manufacturing and construction elements. This is achieved by structuring the cost centres relative to these functions.

A further desirable element of the cost report should be an expenditure forecast, plotting the estimated out-turn costs relative to the project timetable (see 'Expenditure forecasts', below). This will provide useful data to the company's finance department and fund-providers.

On major projects with a substantial contractor element, it is beneficial if the contractor(s) adopts a similar approach to cost reporting, such that the data may be co-ordinated with and integrated into the client's cost reporting system.

Expenditure forecasts

The prime function of expenditure forecasts (sometimes referred to as turnover forecasts) within project cost reports is to provide the company with a prediction of its likely financial outlay, in the form of payments to contractors and suppliers, over the period of the projects(s).

This information will be vital to the company's financial planners in providing the funds to meet the commitments of the capital investment programme. It may also be required by the company's financial backers as a condition of funding arrangements.

Expenditure forecasts may be prepared at each stage of the project estimate, the degree of accuracy increasing with each refinement of the detail of the estimate. The time-scale of the forecast is generally on a monthly basis, as this is the common period for value assessment and account settlement.

The initial forecast may be based on the preliminary cost plan and may be as crude as simple division of the total estimate by the forecast project duration, on a monthly basis.

At the final cost plan stage, the scope of the project works should be fixed, as should the costs. The details of the costs should also be available in a format which is structured to the project execution strategy. The project programme should have been developed to a level whereby the sequence of activities may be related to the cost structure thus determining the spend pattern.

Following project authorization and commencement, further refinement of the forecast may be achieved with the placing of orders (with firm delivery periods) and final negotiation of contractors' fabrication/ installation programmes, with related payment arrangements.

Expenditure forecasts may be displayed in spread sheet format or, more graphically, as 'S' curves. Actual expenditure may be recorded and plotted against the forecast, with any variance being fed back into the Capex funding plans. Also, this variance may be used as a supplementary monitor of project progress, although a margin of error should be allowed for payment/accounting discrepancies.

Change management

Project works, particularly those of significant scale and complexity, or incompleteness of design, are likely to be the subject of some degree of change, both in scope/detail and cost.

This change (usually an increase) may arise from a number of sources including:

☐ Late client changes of requirement.
☐ Development of design detail.
☐ Enhancement of technical specification.
☐ Unforeseen site problems.

The recognition of these changes and their impact upon the out-turn costs of the project is an important feature of cost control. A procedure should be in place which enables the changes to be identified, formalized, costed and fed back into the cost control mechanisms. This procedure is addressed in greater detail below.

Cost engineering

Project cost management is essentially a function and a responsibility of the project manager. However, on major, complex projects a degree of delegation of this responsibility may be required, together with support on information gathering and data preparation.

This support may be provided by:

☐ Cost engineers – to calculate and assemble estimates, budgets and cost centres, to assimilate cost data and prepare cost reports, to identify and investigate cost variance and adverse trends.
☐ Quantity surveyors – to provide commercial, contractual and strategic advice, to prepare estimating data and cost plans, to manage procurement, to prepare cost forecasts, to assess change costs, to value and negotiate contractors' accounts.

The provision of this specialist support is referred to in Chapter 3.

Operation

Assembly of budgets

The progressive stages by which the project budget is prepared is described in Chapter 6.

The initial budget is generally preliminary in nature and may be prepared using the following approximate estimating techniques:

- *Database.* The company may retain its own database of historical project costs from which equivalent or similar cost information may be drawn. This data may be elemental in nature or trade related, for example mechanical plant, electrical installation, building structures, etc. Appropriate adjustment should be made for variance of project situation or complexity and allowance made for inflation. Alternative sources of such data of a more general nature are published by various organizations in both hardback and software format.

- *Unit rates.* Other forms of externally published cost data are based on unit rates. These are generally measurement related and thus require a certain amount of drawn information. They are best suited for works which are relatively simple and/or repetitive in nature, such as pipework, building works, etc. The unit rates may take the form of cost yardsticks, for example cost per square metre for a certain type of finished building. They may also be more detailed in nature, for example cost per meter run of a certain type of pipework. Such rates are generally provided for a wide variety of different situations and are published on an annual basis, which resolves the question of inflation allowance.

- *Factored costs.* This method is suited for projects with a high degree of specialist plant content. Assuming that the supply of these items is predetermined and their costs are known, the balance of the project works may be costed by the application of recognized factors, for example plant costs $\times$ 2·5.

- *Budget quotations.* Contractors and suppliers, particularly those used by the client on a regular basis, will generally be willing to provide budget quotations for relevant sections of the project works. This is, in effect, a 'free' estimating service, although caution should be exercised by the client to avoid any premature commitment to the particular vendor. There should be an appropriate balance between assistance and opportunity. These cost data give greater confidence, as it is a 'market' price, rather than a theoretical one. However, accuracy continues to be determined by the level of detail available. It is also 'flavoured' by the vendor's commercial view of the job potential.

The initial cost plan is developed into the more detailed and accurate final cost plan as and when the design is developed in the form of drawings, schedules and specifications. The availability of these data will not only permit refinement of the initial estimating techniques, but will also provide the information for obtaining fixed price quotations from suppliers and contractors. These quotations should form the backbone of the final cost plan structure as the 'on-site' costs, that is, the cost of the plant, equipment, labour and materials required for the manufacture, fabrication and installation of the project works.

The number and scope of the tender 'packages' is determined by the project contracting strategy (see Chapter 3). The tender process should be conducted competitively and formally (see Chapter 3). When the tenders are received, an appraisal should be undertaken by the project manager and his commercial advisers. A judgement should be made as to which represents the most realistic price for executing the works (which may not be the least expensive). Where competitive tenders are not achievable, prices should be reconciled with alternative forms of estimate.

The selected tender prices may then be assembled, together with estimates for any labour, material or plant required for residual, preparatory or interface works, as the 'on-site' costs. Any works to be undertaken by directly employed labour (other than design, project management and administration) should also be included through internal estimates.

The other major element of the final cost plan is the 'off-site' costs. This will generally include the cost of design, project management, administration and other fees and charges to the project. It should include those costs at all stages of the project, including investigation and other pre-authorization activities. For projects involving process and production plant, decisions will have to be made on the cut-off point for cost commitment once the plant is functional. This would apply to transitional arrangements, such as commissioning, performance trials, operational spares, etc.

Design, project management and other supporting and administrative functions may be undertaken in-house or by consultancy appointments (see 'Consultancy appointments', in Chapter 3). The costs of directly employed or agency staff should be assessed by internal estimate. The costs of consultants should be captured by fixed fee quotations.

Provision should also be made (subject to company accounting policies) for other fees and service charges likely to be incurred. These may include finance and insurance costs, statutory fees, utilities charges, etc.

The final element of the project cost plan is the contingency allowance to cover risk. This is desirable as no form of pre-estimate is 100% inclusive or accurate. The accuracy of the estimate(s) are directly related to the amount of detailed information available which, in turn, is a product of the time and resources committed to it. Risk is not only related to accuracy but also to factors out-with the project's control. This may include external factors such as unpredictable shifts in market places or general inflation. It may also include site-related factors such as unforeseen ground conditions, exceptionally inclement weather, etc.

The contingency allowance may be included as a calculated sum(s) relative to the assessment of the risk, or as a percentage factor of the overall estimate (say 10%), variable with the judgement of the risk factors.

The completed final cost plan is then ready for conversion to the project budget for the implementation phase (see below).

Cost monitoring

The final cost plan used for project authorization becomes the project budget for the purpose of monitoring costs during the project execution phase.

The budget structure is provided by way of the cost plan assembly of on-site and off-site costs. These elements will be broken down into a series of sub-categories which form the basis of cost centres against which incurred costs may be recorded. Further development of the cost centre structure may be required to reflect the project manager's particular strategies for project implementation.

While it is important that the budget structure should match the work plan, it is equally important that costs should be recorded or allocated to the correct cost centre, otherwise misleading comparisons and variances may result. Costs may be incurred in a variety of ways: invoices, valuations, time-sheets, internal transfers, etc. Attention should be paid to the accuracy of referencing and coding of such data at source.

If the procurement process has been employed to establish the cost of contracted work 'packages', this will be followed by the appointment of contractors short-listed from the tendering procedure. This process is likely to account for a substantial portion of the on-site costs. These appointments will be confirmed in orders or formal contracts which will specify payment arrangements which will become project costs.

Major contracts are likely to require progress payments, either on a 'milestone' basis or on a regular monthly basis. Milestone payments are related to the achievement of predetermined activities, usually related to the project programme, and may be expressed as a set sum or as a percentage of the contract price. Monthly progress payments are determined on a 'measure and value' basis and may require the services of a specialist, that is a quantity surveyor, to undertake this assessment and, in an independent capacity, formally certify payment due.

The cost of routine purchases is likely to be captured via normal delivery and invoicing arrangements.

Project accounting procedures are addressed in detail in Chapter 7.

Cost review

Cost monitoring should be undertaken on a continuous basis by the project manager in conjunction with the commercial manager, cost engineers etc. The cost implications of all significant project decisions and events, such as the placing of major orders, should be reviewed at the appropriate time.

Notwithstanding the ongoing monitoring process, the review of project costs should be formalized on a regular basis in the form of a cost report, usually produced monthly. The cost reporting function should be integrated with the overall project reporting procedure (see Chapter 5).

A typical project cost report may include the following sections:

- ☐ *Executive summary/project statement.* The current overall financial status of the project, that is, to budget and expenditure forecast.
- ☐ *Project cost details.* The costs relative to the budget, defined against the project cost centres. This should be subdivided into the current position, the anticipated future view and the forecast final position. Variances between budget and actual cost, and related trends, should be highlighted, as should any departures from the budget structure. Appropriate commentary and explanation should be added. When recording costs, it may be appropriate to differentiate between cost commitments, that is liabilities arising from orders placed, deliveries made, valuations issued, etc., and costs actually paid.
- ☐ *Procurement schedule.* A statement of the financial out-turn of the major contracts awarded and orders placed, relative to the project budget allowances. Variances again should be indicated and commentary added.
- ☐ *Scope change schedule.* A statement of formally instructed changes to the scope of the works (see below), their costs and their financial impact upon the project budget.
- ☐ *Contingency review.* A statement of authorized expenditure of the contingency allowance within the project budget. Purpose of use and allocation to the relevant cost centre should be defined.

Scope changes

An important feature of project cost control is the management of changes in the scope of the project works. These changes may arise from a variety of sources (see Chapter 6) and may emanate from the client or the relevant contractor. However, it is sound commercial practice, as well as being a requirement of most standard forms of contract (see Chapter 3), that the sole authority for change should be the project manager and that changes are formally instructed in writing. In certain contracting arrangements, this authority may be formally delegated to an appointed consulting engineer or architect.

A formal procedure should be established for change mechanisms within the client organization. A proposal in the form of a scope change notice should be made by the relevant party or department, authorized at the appropriate management level and agreed by the project manager. Estimates of the cost and time implications of the change should be made, agreed with the project manager and included in the notice. The scope change document should be signed-off by the relevant parties and then formally instructed to the contractor.

A similar procedure should be adopted for changes instigated by the contractor. For contractual reasons, attention should be paid to the validity and justification of such change proposals. Also, pre-agreement of the cost and time implications with the contractor is desirable.

5. Project management

D. Cole

Introduction

The client's problems with regard to the management of his capital expenditure programme are identified and described in Chapter 1.

Chapter 2 endeavoured to identify client and project management responsibilities, establish practices to develop a capital expenditure plan and provide means of progressing project proposals from inception to authorization.

In this chapter, procedures for project execution, planning, control and reporting are developed using recognized project management practices.

Responsibility for management of project execution and maintenance of procedures lies with the project manager, the client retains overall responsibility for authorization of expenditure variations and sanction of payments.

Project implementation

The implementation phase of a project will commence upon formal authorization by the client in accordance with any established Capex procedures.

Registration of the authorization document confirming a project is active achieves a dual purpose. Recognition of the project in the client's financial systems is established and activation of the projects works, implementation procedures and controls are initiated.

The appointed project manager will now undertake appropriate actions for implementation of project procedures outlined in this chapter. Immediate actions to be considered by the project manager at this point include:

- [ ] finalization and issue of the project scope and procedure document
- [ ] formalization of the project team structure and appointment of personnel
- [ ] establishment of administrative systems
- [ ] commencement of project planning activities
- [ ] establishment of cost control system
- [ ] commencement of procurement activities
- [ ] preparation for construction.

Scope and procedures document

Any project of significant size and/or complexity is required to have a written record of the means of implementation. This record is generally referred to as a 'scope and procedures' document.

The broad purpose of such a document is to define the

- [ ] process/technical scope of the project
- [ ] methods of project implementation
- [ ] procedures and administration associated with implementation
- [ ] roles of personnel required for implementation
- [ ] documentation accompanying implementation.

These matters will have been considered at the investigation stage of the project and a 'draft' scope and procedures document prepared.

The range and content of the scope and procedures document will be dependent upon the size and complexity of the particular project.

As a minimum requirement, the following matters would be defined within the scope and procedures document:

- [ ] *Technical/process scope.* Project definition and objectives, process description, plant identification, services requirements and battery limit conditions, control philosophy, building description and enabling works.
- [ ] *Implementation methodology.* Project management, design responsibility, procurement strategy and work packaging, contractual arrangements, budget control, planning, procurement, construction, safety and handover.
- [ ] *Administration procedures.* Project files, programming, drawing preparation/approval, instructions, variations, expediting, reporting, meetings, payments, costing, quality control, testing and commissioning.
- [ ] *Personnel.* Project management structure, process and engineering disciplines, client liaison, safety management, planning, costing and commercial.
- [ ] *Documentation.* Site regulations, information transmittal, correspondence, site instructions, scope changes/variation orders, expediting inspection reports, payment certificates, internal reports, contractor reports, meeting minutes and takeover certificates.

On large scale (multi-million pound) projects, it may be appropriate for the scope and procedures document to be split into separate documents for the principle elements that is, a technical/process 'specification' and an 'implementation plan' (incorporating procedures, personnel and documentation).

Ancillary documents may also be prepared for particular matters, for example design, planning, cost control, safety, etc.

Implementation procedures

These implementation procedures set out to define the standards to be adopted in the execution of capital expenditure projects from authorization to completion and to provide guidelines for those involved in the management of such projects.

It is anticipated that the procedures will achieve a uniformity and consistency in approach to project execution whilst allowing the flexibility necessary to satisfy individual project requirements.

The guidelines provided within these procedures reflect established technical, commercial and management practices that are in common use.

The implementation phase of a project commences upon formal authorization of the project, the project manager then initiates all appropriate actions for timely and cost effective execution of the project works.

Project set-up (Fig. 1)

During the setting up phase of project implementation the project managers' tasks and responsibilities for execution of the project are established. Management actions regarding project structure, strategy, administration and procedures for procurement, cost control, planning and reporting are addressed at this time.

Immediate actions, including issue of the 'scope and procedures' document, to be undertaken by the project manager are outlined in 'Project implementation' above.

Procedures for project planning and reporting are contained in Chapter 2.

Issue of the 'scope and procedures' document will confirm the agreed strategies for the project and for the commercial and contractual arrangements.

Resources

The structure, organization and resource requirements for the project team will have been identified at the investigation stage. Confirmation of availability of resource personnel will need to be established and any shortfall covered by suitable substitutes, the employment of additional contract personnel or the use of consultants.

A project team structure in graphic form (organagram) identifying team members, disciplines and lines of delegation/communication will be provided by the project manager.

Procurement

A key project management action at the project set-up phase is to proceed with the procurement of goods and services necessary for execution of the

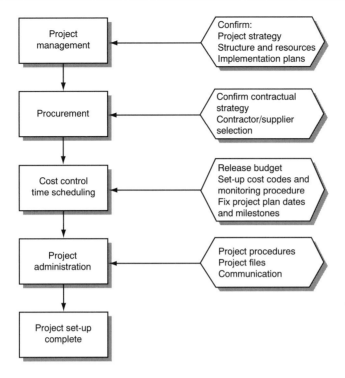

Fig. 1. Project set-up

project. These goods and services will generally have been sourced and priced during the project investigation stage by the issue of enquiries and subsequent bid management process.

In certain circumstances, a project may be authorized on the basis of budget quotations or internal estimates, that is, without fixed prices from external sources. In such circumstances the project manager will instigate procurement of goods and services by means of competitive tendering through the issue of formal enquires.

The effective and timely procurement to meet project budget and programme requirements is the responsibility of the project manager operating within the agreed commercial and contracting strategies.

A full description of the enquiry/tender process, commercial objectives and contracting strategy is set out in 'Commercial Management' (see Chapter 3).

Cost control

The authorization of a project will include the capital expenditure sum budgeted for implementation of the scheme. The sum will have been compiled as a cost plan during investigation and presented in its constituent parts within the authorization documents. This budget break-down, as it becomes following authorization, will provide the yardstick for cost monitoring and control as the project proceeds.

It is the project managers responsibility to execute the project for the predetermined sum. This responsibility includes a general obligation to maintain expenditure within individual budget element constraints. For this purpose the project manager will need to establish systems for budget and expenditure controls and cost monitoring. Such systems would need access to a computerized accounting function that enables day-to-day cost commitments and expenditure arising from orders placed or payments made to be recorded against predetermined, project specific cost centres.

The master control estimate, used to develop the cost plan during investigation, is formatted to reflect the authorized budget elements. As such it is a useful tool to monitor expenditure during project execution and to record revisions to budget elements brought about by change.

Project administration

The requirements for project administration will be set out as admini-stration procedures in the project scope and procedures document. It is the project managers responsibility to develop these procedures to meet the specific needs of a particular project and to maintain all identified documentation.

Project implementation will require a considerable level of communication both within client's internal structure and with external parties (i.e. suppliers, contractors, consultants, statutory organizations, etc.). The scope and procedures document will identify the means, format and method of recording of all communications whether they be undertaken on a formal or informal basis.

Meetings are an essential feature in the management of projects both as an aid to communication and as a record of discussion, conclusion and action. The size, complexity and duration of a project will determine the nature, frequency, agenda and attendance requirements of meetings.

It is likely that a considerable amount of documentation will be accumulated during the life-span of a project. The project leader shall be responsible for establishing and maintaining an orderly and systematic means of retaining and retrieving this information, generally referred to as the 'project files'.

A recommended list of basic files to be established within the 'project file' will be defined in the scope and procedures document. As a general rule all original documentation will be held in the 'project file'. Principal exceptions to this rule would include contract documents and invoice originals. Correspondence and project documents should always bear an identifying project reference.

Responsibility for the maintenance of the project files, arrangements and management of meetings and control/distribution of documents is that of the project manager.

Design and engineering (Fig. 2)

The principles of design for the project works are set out in Chapter 2. These principles would have been considered during the investigation stage and the design responsibility identified.

Process, conceptual and front-end design, based on the submitted design brief, would have been completed at the investigation stage. This design would have formed the basis for enquiry documents aimed at securing fixed prices from vendors and contractors.

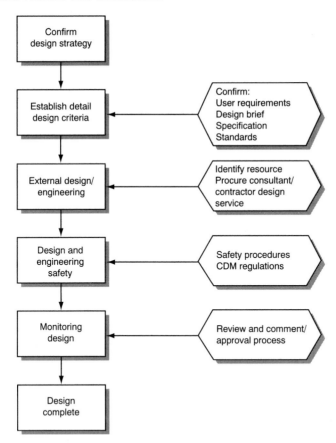

Fig. 2. Design and engineering

This leaves detailed engineering design to be undertaken post-authorization as part of the contractual obligations of selected vendors, suppliers and contractors.

The use of consultants for design at this time is limited to involvement in design of a specialist nature or in an advisory role in the supervision or approval of design.

Prior to the appointment of a consultant the project manager is obliged to provide a comprehensive brief clearly defining: the services to be provided, the tasks to be undertaken, a list of deliverables and a timetable. The consultant will respond by submitting a formal offer setting out his fees and intentions for each stage of design where appropriate.

Engineering

Engineering design is defined as the development of working details that enable manufacture/fabrication/construction to proceed and is comprised of all drawings, sketches, schedules and calculations necessary for this purpose. The responsibility of vendors and contractors to provide such documents will be clearly defined in the contract documents.

It is the responsibility of the project manager to monitor the progress of engineering design. However in areas of speciality this responsibility may be delegated to the appropriate discipline resource assigned to the project team.

The specification and drawings contained in the enquiry documents would have been checked for conformity with the design brief, user requirements and client/British Standards. In turn the contractors offer would have been reviewed against the enquiry specification to ensure that the project proposal met the requirements of the original Capex request.

It is therefore the design and specification included in the contractors offer and subsequently contained in the contract that design deliverables will be monitored against.

Safety in design

During the investigation phase, as part of front-end design, full consideration will have been given to all aspects of project safety.

Incorporated into the design at this stage would have been the requirements of:

☐ clients' company safety manuals, policies and procedures
☐ construction, operational and maintenance safety features
☐ compliance with the Construction (Design and Management) Regulations appropriate to design.

All safety features of the project would have been identified to potential vendors and contractors by way of the enquiry documentation.

At the detail design and engineering stage responsibility for the development of design safety passes to the vendor/contractor.

Design approval

There is a general obligation imposed by model forms of contract for designs prepared by vendors/contractors to be approved by the 'engineer'. The model forms require the engineer to be nominated within the contract documents but permit certain duties of the engineer to be delegated.

A client's policy on the role of the engineer must be clear and unambiguous as must be the intention to delegate specific duties to the project managers.

Chapter 3 discusses the various model forms, their application and the obligations on the part of the client.

The specific duty of approval of designs shall be delegated to the project manager, thereby allowing him to monitor and control this element of the works. The project manager, in turn, may delegate parts of this function to separate engineering disciplines, that is, mechanical, civil, electrical and instrumentation.

In order to maintain an equilibrium between design responsibility it is prudent to restrict the level of input to design approval in the following manner.

'Approval'

Formal approval to be restricted to those matters of design which are fundamental to the client's process or production performance, the operation of plant and the function of services and buildings. Such key matters shall have been developed at front-end design stage and shall form the basic contractual requirements, as defined in the contract specification and drawings.

Thus, any design documentation relating to these key matters and prepared, re-worked or amended by vendors/contractors will require formal submission for approval.

Such documentation should include, but not necessarily be limited to:

☐ process and instrumentation diagrams (P & IDs)
☐ plant general arrangements
☐ floor plans and elevations
☐ services layouts.

'Review and comment'

Design documentation outside the terms of reference for approval, as defined previously, shall nevertheless require scrutinization by the project manager. This procedure is referred to as 'review and comment' and relates to engineering and working design documentation prepared by vendors and contractors.

Such documentation shall incorporate all details required for manufacturing or construction purposes including:

- □ shop/fabrication/assembly drawings
- □ standard component details
- □ structural/architectural details
- □ material/component/plant specifications
- □ quality assurance documents.

The approval/review and comment requirements will be clearly defined in the contract specification, together with procedures and timing for document preparation and submission.

Approval or comment will only be issued when the project manager is satisfied that submitted designs and drawings conform to the technical requirements of the contract. When issued, approvals and comments will need to bear a statement that approval is without prejudice to vendor/contractor obligations under the contract.

Project execution (Fig. 3)

The project phase covers a sequence of events consisting of procurement, manufacture, construction, installation and testing activities. The major project management role during this stage is to monitor progress, costs and quality against the project programme, budget and specification. Co-ordination of off-site activities, interface with client's operations and supervision of the site works are recognized as prime project management functions.

Procurement

At conclusion of project set-up, vendors and contractors will have been identified for the supply of goods and services by way of a bid management process.

Selection of vendors/contractors for the defined work packages or contracts is the responsibility of the project manager and will be made following discussion with the appropriate commercial/contractual resource within the project team. The selection process will involve negotiation with the recommended suppliers to conclude the contract conditions, price, programme and specification.

The procedures associated with bid management, supplier selection, award of contracts and placing of purchase orders are fully described in Chapter 3.

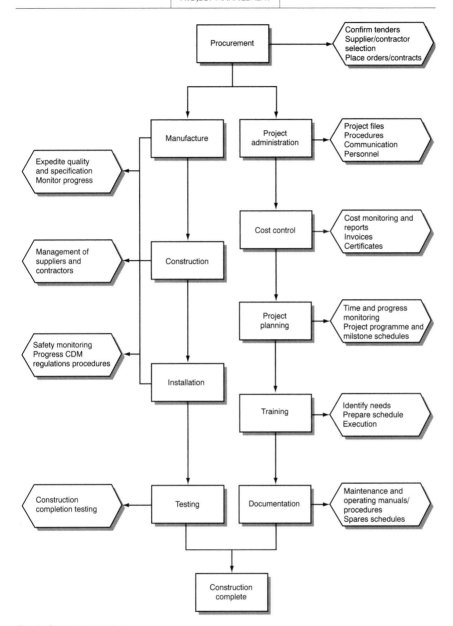

Fig. 3. Project execution

Manufacture

Manufacture can be defined as the off-site activities of a vendor/supplier covering the purchase of materials, their fabrication and assembly into project components.

Off-site activities may form a substantial part of the project works and it is therefore essential that the project manager ensures components are manufactured tested and delivered in accordance with the drawings, specification and programme.

Expediting is an extension of the procurement activity whereby the project manager will monitor off-site manufacture by physical examination of components arising from visits to the suppliers works.

The requirements for expediting, are determined by the project manager to suit the particular mode of supply and possibly the procurement arrangements between the vendor/contractor and his suppliers or sub contractors. Arrangements and timing for expediting visits will be made directly with vendors or in the case of subcontracted supplies, in conjunction with the main contractor. The requirement for expediting visits, their nature and timing, should be clearly specified in the contract or purchase order documents.

The project manager may delegate expediting or testing activities to particular discipline resources where appropriate to the component under manufacture. External agencies may be engaged to carry out independent statutory or certificable testing of specialist components, for example pressure vessels.

At key stages of manufacture or upon completion of component assembly, any specified testing will be implemented by the vendor and witnessed by the project manager. The vendor is required to make available quality assurance procedures, associated documents, material certificates and provide the project manager with all relevant test results.

The project manager will be responsible for provision of a written record of all expediting visits and witnessed tests, preferably on a standard form designed for the purpose.

Assessment of progress of off-site works is to be incorporated into project reporting procedures. Measurements of manufacturing progress against programme and achievements against key milestone dates may be payment related. In these circumstances the project manager will ensure that appropriate documentation is provided in accordance with the contract terms.

Construction

For projects requiring site construction and installation works the activities will commence with the establishment of the 'site'.

The implementation of any such projects is likely to require the presence of external companies on the client's premises in the form of visitors, operatives, delivery vehicles and site works.

Interruption to the client's business or production must be avoided and it is obligatory for the project manager to ensure that project implementation does not interfere with these activities. All project site works by external companies or suppliers must be carefully co-ordinated to meet this obligation.

The extent of arrangements to accommodate the project activities will vary according to the project complexity. It is encumbent upon the project manager to liaise with the client and external companies to establish the arrangements.

During the construction phase it is essential that the scope of the project site works is closely defined to all parties. The contractors obligations are set out in the contract drawings, specifications and predetermined programme.

The timing and extent of the works need to be identified to the client and areas of interface with business activities exposed. It is essential that the project manager gives his full attention to these interfaces to ensure that project objectives are achieved without interruption to the client's production.

The location of the project site would have been identified during project investigation and may be broadly categorized as either

☐ located within an existing building, adjacent to plant or services that may be operational or

☐ located in an open site cleared of plant and services.

Subject to the nature of the particular location the project site will be 'established' by:

☐ Existing building: wherever practical, physical isolation of the new works from existing plant by the erection of temporary screens or the construction of new division walls.

☐ New building/structure: erection of temporary hoardings to the boundary of the site.

Description of specific site conditions and working arrangements would have been communicated to vendors and contractors through the contract specification set out in the enquiry documents during investigation and subsequently contained within the contract.

It is advisable that the project manager, on behalf of the client, provides a set of site regulations for contractors that relates to all anticipated contractor activities within the client's premises. The document, on issue to each contractor, requires a signature of acceptance of the regulations and is a useful tool for the administration of contractors' personnel.

Establishing and maintaining control of project site activities is a prime function of the project manager during construction. Three principal areas of control need to be addressed, management of resources, quality of supply and safety.

The management of resources relates to control of manpower, plant and materials involved in day-to-day site activities. In general terms it is a matter of organization and administration to ensure that resources are appropriately applied to achieve project objectives and programme.

The project manager is responsible for completing project construction and installation to meet specific quality standards. These standards would have been included in the enquiry documents and replicated in the project contract. The contractor is obliged to maintain quality standards for the construction works but responsibility for supervision of the contractor's work lies with the project manager. Contractually this supervision is a delegated responsibility and may be undertaken by specific discipline members of the project team.

It will be the responsibility of the project manager to ensure incorporation of all of the client's safety policies and procedures into the scope of the project execution. Particular attention will be paid to the requirements of the Construction (Design and Management) Regulations with respect to construction activities. Development of the safety plan and the transfer of specific roles and responsibilities to the contractor are areas requiring special consideration.

Additionally general and project specific safety procedures will be determined and agreed with the contractor to reflect

- [ ] the contractor's own safety policies
- [ ] method statements
- [ ] communication to operatives (including subcontractors)
- [ ] safety inspections and reporting.
- [ ] statutory requirements, for example asbestos removal, waste disposal, etc.
- [ ] permit systems.
- [ ] protective and emergency measures.

Inspection and testing

While contractors are obliged by the contract to provide goods and services of specified quality, the project manager will be responsible for establishing procedures for monitoring this undertaking.

Inspections will be carried out during the execution of the construction works as will testing of completed elements of work at predetermining points in the installation programme. Regular site inspections by the project manager or a delegated member of the project team will be made to determine the quality of work in progress. The purpose of the inspection is to identify and report on works failing to meet specification and to initiate a process for corrective action.

As the construction/installation works near completion the inspection and testing of the plant and installation will commence. The inspection requires an orderly and systematic physical examination of the works to confirm compliance with the specification. Such inspections will be carried out jointly by the project manager or his delegated representative and the contractor.

The inspections will need to be carried out to an agreed sequence and timetable to enable them to be implemented on a progressive and timely basis. The timing of the inspections is significant and the advancement of the works to an appropriate state of completion is the responsibility of the contractor. It is advisable that checklists shall be prepared in advance and utilized to record compliance with or failure to meet the specification requirements.

Administration

Throughout the execution phase the project manager will continue to monitor and report on project progress, quality and budget.

Regular inspections of the work in progress will enable measures of completeness achieved to be set against the project plan and an assessment of project progress made. The effects of any slippage against the construction programme will be constantly reviewed to assess the implications to overall project completion and final costs.

Compliance with the specification of the installed works will be overseen and any claimed need for variations to maintain the project scope scrutinised.

Cost control procedures for handling changes in scope, variations to contract, valuations, invoice verification and progress payments are essential to the project manager for the monitoring, reporting and controlling of the project budget.

Such procedures will vary depending on the type of contract or model form of contract employed. Forms of contract and cost control procedures are discussed in Chapters 3 and 6, respectively.

The project manager will communicate the current status of the project by holding regular meetings with both the contractor and the client. Requirements for frequency, format and attendance of project meetings is set out in the scope and procedures document.

Project status will also be communicated via monthly project progress reports. Requirements for project reporting are described in 'Project reports', below.

Training

The training of the client's operating and maintenance personnel will normally be achieved in two phases. Plant/equipment familiarization training will be undertaken at a suitable point towards the completion of installation, when the appropriate plant and main services are available.

The project manager will be responsible for the close co-ordination of training with the contractors activities and must consider safety to be a prime concern.

The second phase of training is for 'hands on' operation for live plant and is closely associated with the proposals for plant commissioning and timing of availability of material resources.

However, both phases of training require the project manager to liaise with the appropriate client business area to fully consider the identification of training needs, the scheduling of training, personnel involved and procedure for executing the training.

Documentation

The levels of documentation to be provided by the contractor and the timing of its presentation will be set out in the contract schedules.

Before completion of the construction phase certain documents, some of which may be in draft form, will be required by the project manager to enable progress of inspection, testing, training and the purchase of long delivery spares to be achieved.

Documents required at this stage include but will not necessarily be limited to

- general arrangement drawings (latest revisions)
- control diagrams (latest revision)
- services diagrams (latest revisions)
- maintenance manuals (draft)
- operating manuals (draft)
- operating procedures (latest revision)
- spares schedules (draft)
- vendor equipment manuals.

The project manager is responsible for the receipt, recording and distribution to appropriate areas of all submitted documentation.

Commissioning and completion (Fig. 4)

Project completion is to be regarded in two distinct phases – (1) the contractual completion associated with the project works undertaken by contractors and suppliers and (2) the handover of the project works to the client's particular operational unit or business area.

The first phase requires the project contractors to meet contractual obligations in respect of the completeness of their works. These obligations will have been defined in general terms in the conditions of contract and shall be set out in detail in the contract specification.

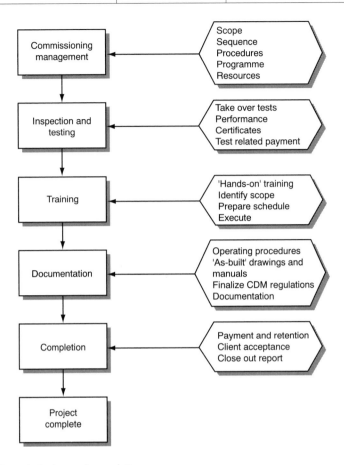

Fig. 4. Commissioning and completion

In addition to achieving construction completion, the contractor may also be obliged to demonstrate functional completion, by way of testing, and thereafter to demonstrate operational effectiveness by way of commissioning and performance testing.

At a predetermined point in the testing, commissioning and performance sequence, responsibility for the completed project works will pass from the contractor to the client. This point is generally referred to as 'takeover'.

The project team now progress to the second phase, that is, final completion of all elements of the works to the satisfaction of the relevant client business area.

The project will then be deemed to be complete in all respects, other than a continuing obligation to correct any latent defects. The closing action of the project manager will be the preparation of the final documentation.

Project completion will be achieved by a sequence of events defined as:

☐ Plant constructed/installed – final inspection, testing and takeover.
☐ After takeover – commissioning and performance testing.
☐ Plant fully operational – handover to client.
☐ Project conclusion final documentation.

Inspection and testing

The project manager is responsible for establishing quality control procedures to monitor the contractors undertaking to provide goods and services to meet the contract specification.

Inspection and testing are an integral part of quality procedures which operate during the execution of the project. At completion these procedures are generally categorized as:

☐ Final inspection and testing which takes place as site works come close to completion. These activities relate to the fulfilment of contractual obligations and will thus be conducted on a formal basis, concluding in takeover.
☐ After takeover on projects involving process or production plant, further inspection and testing will be required in the form of commissioning and performance trials, concluding in handover to the client.

Takeover

Takeover is a contractual term, defined as the formal acceptance of the contract works by the client as the purchaser. It is legally significant as it represents acknowledgement of general completion by the contractor and thereby activates other contractual undertakings, for example release from time penalty, transfer of insurance obligations, payment terms, etc.

The takeover point is generally a single date, predetermined in the contract. For operational purposes, phased or sectional takeover may be adopted. While the takeover date(s) may be defined in the contract, standard forms of contract also provide the option or earlier use or occupation of individual parts of the works, referred to as partial possession.

Takeover is the culmination of a series of procedures with related documentation which are defined hereinafter.

Pre-takeover inspection/tests

As the contractors works near completion inspections and tests, sometimes named as tests on construction completion, are implemented. Such requirements and their sequence have been discussed in the preceding section on procedures for project execution.

The objective of inspection shall be to denote that which is unsatisfactory, referred to as *snagging*. A record of such items shall be made on a standard form or checklist, identifying the item as defective, incomplete, inoperative, etc., as appropriate. The inspection documentation will also denote the action to be taken and the proposed timing, for the purpose of follow-up inspection.

Takeover tests

At a suitable point in the advancement of the works, the final testing of the plant/installation shall commence. In contractual terms, these are referred to as 'tests on completion', that is demonstration that the plant/installation functions in the manner specified.

The scope of such tests and the associated procedures and documentation shall be defined in the contract specification.

Subject to the nature of the particular project, takeover tests may be developed as:

- ☐ final installation tests
- ☐ functional tests
- ☐ precommissioning.

Final installation tests shall relate to the various items of equipment, process, mechanical and electrical services, instrumentation and controls which make up the whole of the plant. The basic objective of final tests will be to determine that all installations are sound and capable of performing to their designed functions prior to energization and operation of the plant. The technical scope and procedure for final test will have been set out in the detailed specifications for the equipment and services.

Functional tests shall relate to the various operational parts of the project including plant, equipment, services, controls and instruments. The purpose of such tests shall to be demonstrate that each individual part operates in the basic manner specified, for example valves open and close, motors turn, etc. These tests shall be relatively simple and shall not require any input of raw materials or product, other than to avoid damage or safety hazard. Temporary or permanent energization may be required for functional tests.

Precommissioning shall relate to the energised operation of the project plant to demonstrate that the various elements perform their basic designed function, for example pumps circulate liquid, conveyors carry materials, etc. If such demonstrations are undertaken without any material input, this may be referred to as 'dry' commissioning.

Subject to the nature of the particular project, the introduction of materials (not necessarily product materials) may be required for precommissioning.

For projects with major plant installations, successful precommissioning using non-product materials (or limited product input) shall be sufficient for takeover purposes.

Takeover certificate

When it is considered that snagging items have been remedied and all takeover tests successfully completed, the works shall be re-inspected and, if deemed satisfactory, a takeover certificate shall be issued. The takeover certificate is a pre-printed form and shall be prepared by the project manager and signed by the client's nominee acting in his capacity as the engineer named in the contract.

In principle, all snagging items shall be cleared prior to the issue of takeover certificate. However, minor items may be permitted to remain outstanding and incomplete, provided that the overall operation or any particular function of the project is not affected. Such items shall be listed as exceptions or exclusions on the takeover certificate. The inspection record documentation identifying these items may also be appended to the certificate.

In any event, the list of exceptions or exclusions shall not be extensive.

The takeover certificate is a contractual document for use with external parties. A similar form can be developed and used for internal 'handover' and the means to record the client's acceptance of the project.

Commissioning

For projects involving process or production plant there is a requirement for the Project Team to set the plant to work with product materials, operating at its desired output, utilizing its designed functions, that is commissioning.

This is a key element of project implementation and the project manager shall ensure that appropriate allowance is made in time and resources.The project manager is responsible for the identification and definition of the commissioning scope and any appropriate procedures.

A commissioning team shall be assembled using engineers and technicians of the appropriate discipline, including any necessary resources from the client's operations.

Commissioning will proceed to a predetermined sequence and timetable, the objective being to handover the fully operational plant by an agreed date.

Commissioning shall require the operation, monitoring and adjustment of the plant using product, often referred to as *wet* commissioning. The project manager shall liaise with the appropriate client production personnel to plan, co-ordinate and control this activity.

An extension of the commissioning function may be an obligation for specialist plant contractors to carry out trials of the functional plant to verify that the designed output is being achieved, that is, performance tests. This requirement shall arise where the contractor has undertaken the contractual responsibility to provide equipment to a performance specification and usually relates to production units.

Performance tests

The requirement for performance tests shall be carefully defined in the project specification as the achievement (or otherwise) may have financial consequences. It is therefore essential that the following elements are fully described:

☐ number, timing and duration of tests
☐ operational conditions and personnel (client/contractor)
☐ performance criteria (for measurement/assessment purposes)
☐ specification of input/output materials
☐ identification of variants (machine down-time, etc.)
☐ witnessing and recording arrangements
☐ recording documentation.

Upon the successful completion of commissioning and performance trials, the plant shall be handed over to the client business area or appropriate operational department. The handover procedure can require the use of an internal 'takeover certificate', to be offered by the project manager and acknowledged by the client.

Operations and maintenance

For projects involving process or production plant, as soon as that plant is operational, there shall be a requirement for project management liaison with the client's operating areas.

This liaison will facilitate the following requirements:

☐ to implement, monitor and control commissioning activities and performance trials as previously described
☐ to train the staff of the client's end-user in the operation of the new plant
☐ to familiarize engineering staff with the functions of the new plant for maintenance purposes.

With regard to the training of personnel it shall be an obligation of the project manager to provide for the client's use an appropriate number of sets of plant operation and maintenance manuals. These manuals shall incorporate all necessary technical data from suppliers and contractors involved in the project.

Additionally, as-built drawings, schedules and services diagrams relevant to project are to be made available for the client's records. The requirements for suppliers and contractors to provide operation, maintenance and as-built data shall be defined in the contract specifications.

Operator training may be conducted on an informal basis during the commissioning period. For large-scale projects or specialist plant and equipment, it is advisable for formal training sessions to be conducted by the principle contractor or specialist.

These training sessions and their objectives shall be scheduled by the project manager to fully meet the client's needs and time constraints. The requirement for contractors and suppliers to provide this service shall be defined in contract specifications.

Project conclusion

Project completion is defined as the date on which 'commissioning is finalized'. However, there are continuing responsibilities for the project manager beyond that date.

Upon completion of the project and handover to the client, the final action of the project manager shall be the preparation of the project close-out report.

The purpose of this report shall be to provide a formal record of the project for future reference. The style and content of the report is set out in 'Project reporting and documentation', below.

The contractual responsibilities of suppliers and contractors shall continue after project completion. Standard forms of contract incorporate a 'warranty' period, also referred to as the 'maintenance' or 'defects' liability period. This period shall generally be of 12 months duration from the date of takeover.

During this period the supplier/contractor shall be obliged to return to site to correct or replace any defective works. At the conclusion of this period, the project manager shall ensure that any outstanding defects have been remedied. If appropriate, a final inspection shall be arranged with the contractor.

When the project manager is satisfied that no defects remain a final certificate shall be issued to the contractor.

Project planning

It is a necessary pre-requisite for all projects that the works be implemented to a planned and co-ordinated timetable confirmed in the form of a project programme.

Timetable is an important factor in the development of a project and consideration should be given at the earliest opportunity, that is, during the investigation stage. When considering the project strategy the project leader will review the timing requirements and prepare a preliminary plan. The style and content of the preliminary project programme is dependent upon the nature and scale of the particular project.

Upon authorization of the project, the project leader will develop the preliminary programme into the working project programme. For projects of larger scale and complexity, the detailed development of the programme will need to be undertaken by a planning engineer. The project programme is considered to be the master programme for all planning and reporting activities. Project activities are to be subdivided into those to be undertaken by external sources (vendors and contractors) and those to be undertaken by the project team.

Subject to the size and complexity of the project, subsidiary programmes may be prepared for particular project activities. These programmes are generally likely to relate to the activities of vendors and contractors who will be required to prepare the detailed, working programmes accordingly. All programmes prepared by external sources shall be co-ordinated with the project programme. Separate programmes will be developed for internal team controlled activities, for example enabling works, services tie-ins, plant testing and commissioning. These programmes will need to be co-ordinated with the related vendor/contractor activity programmes.

Effective planning, progress monitoring and progress reporting are vital project management tools. The establishment of a plan and the communication of planning information to the parties involved facilitates implementation of the work. Progress monitoring and reporting allows this work to be controlled and directed in a manner which ensures that a project is completed to specification, on time and within a budget.

A plan may consist of a specified start and finish date or a series of start and finish dates for various phases or features of work. More complex schedules show the planned sequence of work in sufficient detail to illustrate how the project will be accomplished. Sophistication may range from a simple bar chart to a detailed logic network. A project plan when properly developed, provides the user with an effective management tool to control the critical aspects of performance.

During practical use, the plan becomes an early warning device for identifying situations where problems are developing but may be avoided with proper management attention and action.

Planning objectives

The planning objectives of each project will differ depending upon the complexity or level or information required. Where appropriate, identification of planning objectives for a project will result from discussions between the project manager and a planning engineer and cover the following topics:

- ☐ *Planning support*: to agree the level of planning support required for each investigation/project and to establish with the project manager and discipline engineers the full extent of the work.
- ☐ *Establish important dates*: list out the main activities and establish the flow of action from initiation to project completion; establish dates at which blocks of activities start and finish (milestone dates); fix dates at which major events will occur (target dates); draw out milestone reports for monitoring and reporting purposes; draw out summary bar charts showing periods of time for major sections of work.
- ☐ *Detailed programming*: list activities in a logical sequence, establish durations, dependencies and connect each item of work into the overall plan (network programme).
- ☐ *Resourcing and cashflow*: produce programming information for in-house resourcing, develop a cashflow programme showing estimated expenditure during the course of the project.
- ☐ *Monitoring and progress measurement*: set up methods to measure progress for reporting purposes; assist with the expedition of orders from placement to delivery through procurement schedules; monitor design progress through drawing schedules.
- ☐ *Monitoring of contractors reports*: ensure programming and progress reporting procedures included in tender documentation are adhered to during the contract; ensure that contractors use acceptable routines relating to programming and progress reporting; monitor contractors project reports and progress against programme; monitor contractors resources.
- ☐ *Progress and reporting details*: analyse progress information and review forecast milestones; highlight trends and variances that initiate corrective action; attend project review meetings; produce progress reports as required.

Methodology

The planning of a project can be divided into three distinct phases:

- ☐ Project identification: at which stage a project is described on the basis of general technical data and an initial profitability analysis.
- ☐ Project definition: where the details of a project are refined more precisely reviewing in particular, design parameters, estimated cost, and target dates.
- ☐ Project execution: which is the stage of detailed planning where a

project is taken through design, procurement and construction to final commissioning and handover to production.

This methodology describes the planning levels and documentation used during the investigation (identification and definition) and project execution phases.

Level 1 milestone report

The first level of planning and documentation is a milestone report.

The report is designed to enable monitoring of progress on all investigations and projects, current to the Capex plan, against three key dates. These milestone dates identify: investigation start, project authorization and project completion.

Each milestone is listed with a planned, latest forecast and actual (when achieved) set of dates. Forecast dates are regularly reviewed by the individual project managers and the report revised where necessary.

This level of reporting provides a simple planned and forecast time frame for investigations/projects and is used to monitor progress against the annual operating plan (AOP) and to assist in management of resources.

Level 2 summary programme

Summary planning provides for planning of projects by the project manager through the use of manually drawn bar charts. A summary programme would normally be drawn for a small project or applied to any project when trying to assess a critical milestone or key date during an investigation. The system would normally be used on projects which have a clearly defined scope and a small number of easily defined activities and contract packages. This programme is issued in bar chart form and progress is shown as a percentage of each activity bar against a 'time now' line.

Level 3 detailed programme

The third level of planning and documentation covers the planning of projects at a detailed level. The project is broken down into a list of activities that describe the works from start to completion. The activities are arranged sequentially, given an estimated time duration and drawn in a network form.

A detailed programme would normally be provided for a large project or one expected to be particularly complicated. The programme and documentation would include:

☐ a critical path network
☐ programme bar chart
☐ progress curve

- ☐ resource histogram
- ☐ progress monitoring and reporting systems.

Prior to preparation of the programme the project manager will discuss and agree with the planner the extent of detail required. The development of the planning documentation then passes through four distinct stages:

- ☐ preparation and establishment of the programme
- ☐ setting up of the progress measurement system
- ☐ measurement of progress achieved
- ☐ analysis of the data collected and preparation of progress reports.

These processes are described in detail in the following section on planning procedures.

Planning procedures

These procedures have been developed to guide and assist the project manager to set up, administer, control and report all project planning activities from inception to completion.

The procedures outline the planning requirements associated with a project, they should be used as guidelines to follow and be applied in a consistent manner.

This section assists the key planning elements in:

- ☐ the setting up of a plan
- ☐ the agreement of project milestones
- ☐ confirmation of the management organization
- ☐ development of programmes and schedules
- ☐ establishment of the progress measurement and control system
- ☐ clarification of a reporting system.

The following procedures incorporate proven systems and techniques which are common practice in general industry. They should be suitably interpreted to meet the requirements of each individual project.

Preparation of the project programme

After discussing the scope of work with the project team, the planner will prepare a list of activities in logical order of precedence (a logic-linked network). Each activity will show its inter-dependency, its duration and if required its resource requirement.

The network will normally identify all activities covering design, approvals, procurement, subcontracts, manufacture/fabrication, site mobilization, site construction/installation, testing/commissioning and handover.

Networks

The use of networks in planning, offers the project team a powerful method of controlling and guiding the project. In particular the principles of networking provides a basis for:

☐ describing the activities of a project
☐ monitoring and controlling progress
☐ calculating critical and sub-critical courses of action
☐ identifying those activities which have float time and for evaluating the float
☐ assessing the effects of change to a particular course of action on a project
☐ planning for future action and resourcing
☐ linking activities with other aspects of a project.

The use of computer software adds speed to the calculation of a network and greatly improves the presentation of planning information.

Networks are drawn in precedence format which means that each activity is described in a box and each box is linked within the network by a line. The flow of activities is demonstrated by arrows. When drawing a network it is important to understand the full scope of the project. Most projects can be broken into four distinct stages, namely:

☐ investigation
☐ design
☐ procurement/fabrication
☐ construction/commissioning.

Within these four headings will be incorporated a list of detailed activities and a set of target dates.

Depending on the size and complexity of a project it may be necessary to apply sub-networks. This system allows the user to analyse a small number of activities in more detail after which the results are fed back into the overall network.

The project network forms the basis of progress control and can be used to set up control schedules, activity listings and regular progress reports.

Bar charts

The bar chart is the normal method of displaying programming information and provides a basis for:

☐ listing every activity in a programme, each with start and finish dates shown by a bar
☐ colour or shade coding like activities in order to highlight particular aspects of the programme
☐ graphically highlighting the critical path of a group of activities

- [ ] grouping activities by section, building, discipline, order, process or any other listing that may be appropriate
- [ ] showing graphically the float attached to each activity in a network
- [ ] showing links between succeeding and preceding activities
- [ ] showing progress and highlighting those activities which are ahead or behind a particular 'time now line' or 'cut-off date'
- [ ] demonstrating progress by comparing planned dates with actual dates and inserting forecast dates
- [ ] showing different date arrangements including alternative calendars and week numbers.

Any number of legends, headings, notes, weightings and titles can be added to a bar chart making it one of the most powerful methods of reporting the progress of a project.

Bar charts can be drawn manually but in terms of accuracy it is better to use a network or linked bar chart as the base document. Base information can then be easily transferred by available computer software.

If a network is not used as the base document then it is often difficult to see the full implication of changes.

For most projects two forms of bar chart are used; a detailed bar chart and a summary bar chart.

The detailed bar chart is used to list every activity to highlight the critical path and to show completed activities up to a defined cut-off date. This type of presentation is updated on a monthly basis, gives a dynamic presentation and is very useful for the day to day running of a project. Any alteration to the end date of a project is immediately evident and allows corrective action to be taken to minimize the effect of unavoidable delays or changes.

The summary bar chart is drawn to report progress in summary form with the main objective of showing completed activities, highlighting any changes to the base programme which may have occurred and reporting progress at a defined cut-off date. The summary bar chart would usually be based on a detailed network and be included in regular progress reports.

Schedules

The schedule has become a useful document in highlighting actions to be taken and progressing work in the areas of design/drawing and procurement.

Drawing schedule/design report

On most projects the products of the design phase are shown on drawings and documents which are subject to client comment/approval before being completed and issued for construction. The drawing schedule or design report is used to list the drawings/documents to be produced and the dates at which these details are to be issued. The planned dates are taken from the project bar chart and entered on the document schedule. If planned dates are altered the latest forecast dates are shown in brackets.

Procurement schedule

The progress of procurement is listed and monitored by the procurement schedule. Each step of the procurement process from enquiry through to delivery of an item on site is dated from the project bar chart and itemized on the procurement schedule. When planned dates are revised they become forecast dates and the change is recorded on the schedule. The procurement schedule is updated at regular intervals to coincide with the project reporting times.

Typical examples of both drawing/design and procurement schedules are contained in 'Documentation', below.

Progress measurement

Having established the programme of work in the form of a network or bar chart, it is necessary to have a progress measurement system.

The measurement system establishes a detailed level of progress calculation and ultimately creates an effective means of programme control and reporting. By weighing each network or bar chart activity as proportioned elements of the whole programme and then summing the elements against activity time durations, a planned progress system is established.

The planned progress is ultimately expressed in graphical form as a progress curve.

Physical or actual progress is expressed by summing the proportionate values of work achieved by a measured activity element against actual time durations.

Measurement of design is made against identified deliverables, that is, drawings, data sheets, specifications or calculations.

Procurement of goods or services is measured against tenders issued, bids reviewed, contracts/orders placed, manufacture, delivery and/or site commencement.

For construction, measurement is by quantitive assessment of completed work, for example, tonnes of steel erected, metres of pipework installed, or against agreed percentages of achievement in activity completion.

Payments to contractors on physical progress can be linked with this system and the measurement calculation may be co-ordinated through a quantity surveyor.

Analysis of the data, by comparing the planned and actual progress calculations, is used to assess overall progress status, contractor performance, trends and for decisions on any corrective action deemed to be necessary.

Contractor programmes

Working programmes from external sources, such as vendors and contractors, are to be prepared to reflect the contracting strategy established for the particular project.

Each vendor and contractor of significant size will be obliged to prepare a detailed programme to reflect their contractual undertakings. This obligation must be clearly defined and agreed at procurement stage.

The number and scope of such programmes shall be determined by the final work packaging and procurement arrangements. The project leader shall ensure that all vendor/contractor programmes comply with and are incorporated into the master project programme.

The scope of vendor/contractor programmes will reflect the nature and scale of their particular operations, but as a minimum requirement, shall incorporate the following activities:

Vendors (related to each major item of supply):
- [ ] design/drawing preparation
- [ ] material/component procurement
- [ ] manufacturing stages
- [ ] site delivery.

Contractors (related to each contract/work package):
- [ ] design/drawing preparation
- [ ] subcontractor/supplier procurement
- [ ] off-site fabrication
- [ ] site delivery
- [ ] site construction
- [ ] test and commission.

The general requirements on contractors for planning, progress measurements and reporting documentation will reflect the value scope and complexity of the contracts or work packages.

The following is intended as a general guide and describes the planning and documentation requirements for a relatively high value and complex project needing a high level of control across all phases of the contract. Requirements for more easily defined projects of lower complexity and value would be reviewed as part of the contracting strategy.

Tender programme

The contractor shall prepare and submit a tender programme for the works.

The programme shall be compiled by means of critical path analysis and submitted in bar chart form that summarizes the network.

The tender programme shall show all significant milestone dates including

- [ ] award of contract
- [ ] commencement/completion of design
- [ ] procurement and deliveries
- [ ] completion of construction/installation
- [ ] testing and overall completion
- [ ] handover.

A tender programme network will typically comprise activities covering design, approvals, procurement, manufacture, delivery, construction, testing and commissioning.

The contractor is also to prepare and submit a table of resources necessary to achieve the programme. This table shall be in the form of a histogram and list each discipline/trade.

The contractor shall further prepare and submit an overall contract progress curve with attendant progress calculations.

Outline method statements for major construction and installation activities will be included.

Contract programme

During the contract phase the planning information will comprise

- [ ] a critical path network with a time analysis showing planned and actual start and finish dates, durations and floats for each activity. This information may be provided on a detailed bar chart
- [ ] a summary contract bar chart based on the network. The bar chart shall contain a minimum of 50 activities and will be used as the basis for progress reporting
- [ ] progress curves for design procurement, construction and overall contract
- [ ] site labour histogram showing discipline/trade
- [ ] a 6 week's look ahead activity list
- [ ] a critical activities listing
- [ ] method statements for all principal activities.

The contractor will, in addition, submit a full monthly progress report throughout the duration of the contract. The report shall cover a complete calendar month with a predetermined progress cut-off date and is to be received by the project manager within an agreed period from progress cut-off. A description of the format for contractors progress reports is contained in 'Project reporting', below.

Pre-contract review

During the pre-contract review period, the selected contractor may be required to modify or update the tender programme.

After approval by the project manager, the contractor's programme (network and bar chart), histogram and progress curves shall be included in the contract and used as a basis for the contractor to develop the detailed planning requirements.

Planning systems

The rapid development of computer software for use as a planning tool has resulted in computers playing an increasingly important part in the calculation and presentation of planning information. There are many different computer software packages available designed to assist the function of project planning and to meet a wide range of programming and reporting needs.

The potential of three packages in common use is identified but no attempt or claim is made to promote them or their suitability over other available packages.

Pertmaster Advance

The *Pertmaster Advance* (PMA) software package has been developed to assist the user in most of the exercises required in planning.

In particular PMA covers:

☐ the forward and backward pass calculations of a network
☐ the identification of 'loops' or logic errors
☐ the identification of critical paths
☐ resourcing and resource smoothing
☐ date tabulation
☐ sorting of information
☐ correcting of errors and looking at 'what if' situations
☐ printing of bar charts
☐ printing of network programmes.

Power Project

Power Project has been written around a graphics package. Data are entered to enable a schedule to be drawn directly onto the screen as a linked bar chart rather than the conventional network method. As a programme is updated the links between the bars cause a change to influence the overall programme. *Power Project* offers certain advantages over the more conventional systems in that a user can immediately view the programme in bar chart form. It is more suited for a simple schedule that does not require the development of a logic network. It has a good internal graphics capability and permits the user to improve the graphics once the bar chart has been drawn. It is most suitable for presentation purposes. The package will also produce manpower histograms and progress curves.

Microsoft Project

Microsoft Project is developed as a project management package designed to assist in the establishment and control of the project plan.

Microsoft Project covers:

☐ basic tasks of defining the project, identifying project milestones, assessing resource requirements and creating a realistic project schedule

☐ project management tools, Gantt charts, Pert charts, links and dependencies, critical path, recalculation of schedules and resource levelling

☐ communication of progress against time and cost using a variety of customized reports and information presentations.

Application

Selection of an appropriate software package to suit a particular capital expenditure project plan will be dependent upon the level of project control, reporting and presentation required.

It is important that the selected software is used to its highest potential and that updates to the package are constantly reviewed to maintain efficient levels of utilization.

Planning documentation

Levels of documentation required for the recording of project progress will vary and be determined by the nature, value and scale of the particular project.

The use of typical forms set out in a standardized format will ensure a consistent approach to the control and reporting of project progress.

Sample forms concerning the various documents discussed in 'Planning Procedures' above are illustrated in Figs 5–13.

DOCUMENT/DRAWING PROGRESS CHART

DRAWING No.

ISSUE DATE:

SHEET   OF

REVISION:

CONTRACT

TITLE	DRG No.	EST HRS WTG	START DATE		COMPLETION DATE		% COMPLETE										PRO ACH
			PLANNED	ACTUAL	PLANNED	ACTUAL	10	20	30	40	50	60	70	80	90	100	

Fig. 5. Documenting/drawing progress chart

DRG. No.	DESCRIPTION	ISSUE No.	% PROG.	START DATE PLANNED ACTUAL		ISSUED FOR COMMT. PLANNED ACTUAL		APPROVAL DATE PLANNED ACTUAL		ISSUED FOR CONST. PLANNED ACTUAL	

FORECAST DATES SHOWN THUS (· · · · · · · · · ·)

Fig. 6. Design report

ITEM No.	DESCRIPTION	*	PREP SPEC/ DATA SHT.	ISSUE ENQUIRY	RECEIVE BIDS	EVALUATE AND SELECT	PLACE ORDER	RECEIVE VEND. DETS.	INSPECT/ TEST	DELIVERY EX-WORKS	DELIVERY ON-SITE	REQUIRED ON-SITE
		P										
		F										
		A										
		P										
		F										
		A										
		P										
		F										
		A										
		P										
		F										
		A										
		P										
		F										
		A										
		P										
		F										
		A										
		P										
		F										
		A										
		P										
		F										
		A										

P* = Planned   F = Forecast   A = Actual

Fig. 7. Procurement schedule

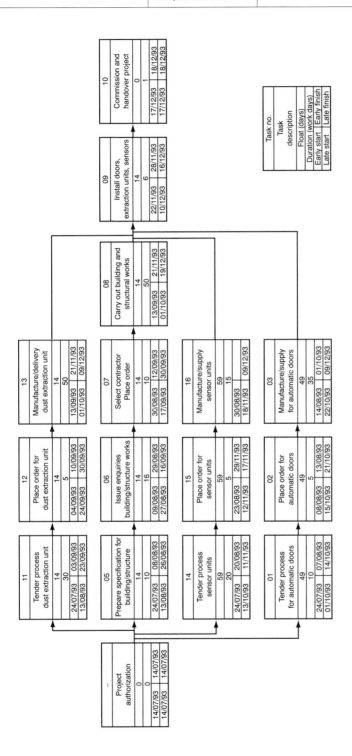

Fig. 8. Programme network

DESCRIPTION	PLANNED		ACTUAL (forecast) *		%P	%A	TIME NOW
	START	FINISH	START	FINISH			YEAR / MONTH / DATE
OVERALL DESIGN	26/06/92	13/10/92	26/06/92	(18/12/92)	100	98	
PROCUREMENT							
Prepare scope and issue enq.	31/01/92	06/03/92	06/01/92	13/03/92	100	100	
Bidding period	09/03/92	03/04/92	16/03/92	13/04/92	100	100	
Evaluate bids/place contract	06/04/92	06/05/92	13/04/92	19/06/92	100	100	
Issue enquiry/order matls	03/07/92	06/10/92	03/07/92	26/11/92	100	100	
Fab/deliver matls to site	08/09/92	17/11/92	08/09/92	(12/02/92)	100	50	
Subconts—enq/place order	13/07/92	05/10/92	13/07/92	25/11/92	100	100	
Site mobilization	02/11/92	06/11/92	23/11/92	25/11/92	100	100	
CONSTRUCTION							
Site installation—structural	09/11/92	02/04/93	26/11/92		10	5	
Site installation—mech/piping	16/11/92	23/04/93	23/11/92		10	5	
Site installation—elec/testing	14/12/92	27/04/93					
Final connections	27/04/93	27/04/93					
COMMISSIONING	28/04/93	24/05/93					

*All forecast dates are the same as planned—unless stated otherwise

LEGEND: ■ = Original plan

PROJECT TITLE

Fig. 9. Summary bar chart

ACTY	DESCRIPTION	PLANNED (1992) Start	Finish
	ABC review/approve pipe rack	06/01	17/01
	Filters. Confirm spec./price/delivery	06/01	24/01
	Complete pipework design	06/01	13/03
	Filters. Obtain vendor details	27/01	31/01
	Prepare outline support details	27/01	07/02
	Tender period for filter units	27/01	26/02
	ABC approve filter unit details	03/02	14/02
	Prepare enquiry for support steelwork	10/02	14/02
	ABC detail filter controls	17/02	28/02
	Tender period for support steelwork	17/02	13/03
	Design interface. Pipework/filters	24/02	27/03
	Order valves for pipe break-ins	28/02	29/02
	Eval. tenders/order filter units	02/03	20/03
	Complete enquiry docs for pipework	09/03	13/03
	Order instruments	16/03	20/03
	Evaluate tenders/order support steel	16/03	27/03
	Tender period for pipework	16/03	03/04
	ABC detail electrics	23/03	03/04
	Manufacture/deliver filter units	23/03	10/07
	Tender process for strainers	23/03	24/04
	Fabricate/deliver steelwork	30/03	08/05
	Evaluate tenders/order pipework	06/04	10/04
	Establish site works	06/04	16/04
	Order materials for pipe break-ins	13/04	16/04
	Prepare pipe shop fab'n dr'gs	23/03	24/04

Fig. 10. Detailed bar chart

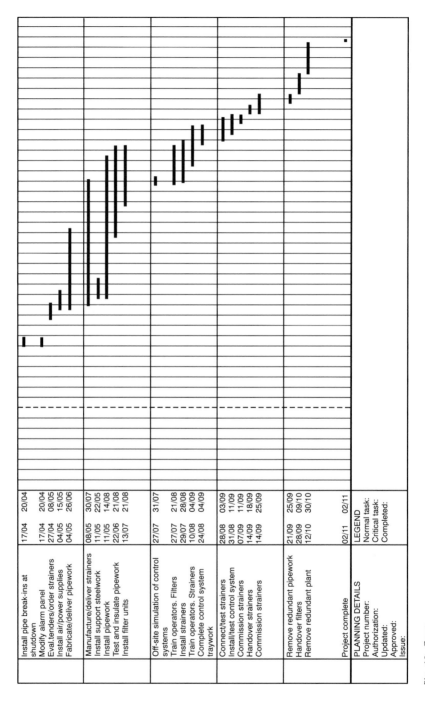

Install pipe break-ins at shutdown	17/04	20/04
Modify alarm panel	17/04	20/04
Eval.tenders/order strainers	27/04	08/05
Install air/power supplies	04/05	15/05
Fabricate/deliver pipework	04/05	26/06
Manufacture/deliver strainers	08/05	30/07
Install support steelwork	11/05	22/05
Install pipework	11/05	14/08
Test and insulate pipework	22/06	21/08
Install filter units	13/07	21/08
Off-site simulation of control systems	27/07	31/07
Train operators. Filters	27/07	21/08
Install strainers	29/07	28/08
Train operators. Strainers	10/08	04/09
Complete control system traywork	24/08	04/09
Connect/test strainers	28/08	03/09
Install/test control system	31/08	11/09
Commission strainers	07/09	11/09
Handover strainers	14/09	18/09
Commission strainers	14/09	25/09
Remove redundant pipework	21/09	25/09
Handover filters	28/09	09/10
Remove redundant plant	12/10	30/10
Project complete	02/11	02/11

PLANNING DETAILS
Project number:
Authorization:
Updated:
Approved:
Issue:

LEGEND
Normal task:
Critical task:
Completed:

Fig. 10. Cont.

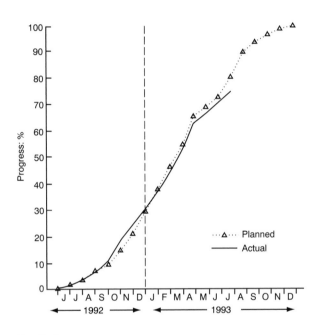

Fig. 11. Progress curves

Cumulative progress: %		
Date	Planned	Actual
30/06/92	1·7	1·7
31/07/92	3·5	3·4
24/08/92	6·6	6·4
30/09/92	9·5	10·3
30/10/92	14·8	18·5
27/11/92	21·1	24·1
31/12/92	29·3	30·7
29/01/93	37·7	36·9
26/02/93	46·0	43·6
26/03/93	53·6	53·0
30/04/93	65·5	62·8
28/06/93	68·9	66·6
25/06/93	72·7	71·3
30/07/93	80·4	74·9
31/08/93	90·0	
30/09/93	93·7	
31/10/93	96·5	
30/11/93	98·7	
31/12/93	100	

	PLANNED		WEIGHTING	% COMPLETE		WEIGHTING ACHIEVED	
	START	FINISH		PLANNED	ACTUAL	PLANNED	ACTUAL
DESIGN:							
TOTAL DESIGN -							
PROCUREMENT:							
TOTAL PROCUREMENT -							
SITE WORKS INCLUDING:							
- TESTING							
- COMMISSIONING							
- TRAINING							
TOTAL SITE WORKS -							
OVERALL PROJECT							

Fig. 12. Progress calculation sheet

ACT	DESCRIPTION	WTG.	% COMP	29-01-93		26-02-93		26-03-93		30-04-93		28-05-93		25-06-93		30-07-93	
				Plan	Actual	Plan	Actual	Plan	Actual	Plan	Actual	Plan	Actual	Plan	Actual	Plan	Actual
	MIXER	-	-														
005	COMPLETE F/E DES/ISSUE ENQ.	0.50	100%	0.50	0.50	0.50	0.50	0.50	0.50	0.50	0.50	0.50	0.50	0.50	0.50	0.50	0.50
015	EVALUATE BIDS & SELECT	0.25	100%	0.25	0.25	0.25	0.25	0.25	0.25	0.25	0.25	0.25	0.25	0.25	0.25	0.25	0.25
020	PLACE ORDER	0.25	100%	0.25	0.25	0.25	0.25	0.25	0.25	0.25	0.25	0.25	0.25		0.25	0.25	0.25
025	MANUFACTURE/DELIVER	7.00	100%	3.20	4.20	4.90	4.90	5.95	6.65	7.00	6.65	7.00	7.00	7.00	7.00	7.00	7.00
	PUMP	-	-														
030	COMPLETE F/E DES/ISSUE ENQ.	0.50	100%	0.50	0.50	0.50	0.50	0.50	0.50	0.50	0.50	0.50	0.50	0.50	0.50	0.50	0.50
040	EVALUATE BIDS & SELECT	0.25	100%	0.25	0.25	0.25	0.25	0.25	0.25	0.25	0.25	0.25	0.25	0.25	0.25	0.25	0.25
045	PLACE ORDER	0.25	100%	0.25	0.25	0.25	0.25	0.25	0.25	0.25	0.25	0.25	0.25	0.25	0.25	0.25	0.25
050	MANUFACTURE/DELIVER	2.00	100%	1.20	1.60	1.80	1.80	2.00	2.00	2.00	2.00	2.00	2.00	2.00	2.00	2.00	2.00
	CONDENSER	-	-														
055	COMPLETE F/E DES/ISSUE ENQ.	0.50	100%	0.50	0.50	0.50	0.50	0.50	0.50	0.50	0.50	0.50	0.50	0.50	0.50	0.50	0.50
065	EVALUATE BIDS & SELECT	0.25	100%	0.25	0.25	0.25	0.25	0.25	0.25	0.25	0.25	0.25	0.25	0.25	0.25	0.25	0.25
070	PLACE ORDER	0.25	100%	0.25	0.25	0.25	0.25	0.25	0.25	0.25	0.25	0.25	0.25	0.25	0.25	0.25	0.25
075	MANUFACTURE/DELIVER	9.00	100%	4.00	5.55	5.55	6.75	7.65	9.00	9.00	9.00	9.00	9.00	9.00	9.00	9.00	9.00
080	COMPLETE F/E DES/ISSUE ENQ.	0.50	100%	0.50	0.50	0.50	0.50	0.50	0.50	0.50	0.50	0.50	0.50	0.50	0.50	0.50	0.50
090	EVALUATE BIDS & SELECT	0.25	100%	0.25	0.25	0.25	0.25	0.25	0.25	0.25	0.25	0.25	0.25	0.25	0.25	0.25	0.25
095	PLACE ORDER	0.25	100%	0.25	0.25	0.25	0.25	0.25	0.25	0.25	0.25	0.25	0.25	0.25	0.25	0.25	0.25
100	MANUFACTURE/DELIVER	4.00	100%	4.00	3.20	3.40	3.60	4.00	4.00	4.00	4.00	4.00	4.00	4.00	4.00	4.00	4.00
	TANKS	-	-														
105	COMPLETE F/E DES/ISSUE ENQ.	0.50	100%	0.50	0.50	0.50	0.50	0.50	0.50	0.50	0.50	0.50	0.50	0.50	0.50	0.50	0.50
115	EVALUATE BIDS & SELECT	0.25	100%	0.25	0.25	0.25	0.25	0.25	0.25	0.25	0.25	0.25	0.25	0.25	0.25	0.25	0.25
120	PLACE ORDER	0.25	100%	0.25	0.25	0.25	0.25	0.25	0.25	0.25	0.25	0.25	0.25	0.25	0.25	0.25	0.25
125	MANUFACTURE/DELIVER	2.00	100%	2.00	1.60	1.70	1.80	2.00	2.00	2.00	2.00	2.00	2.00	2.00	2.00	2.00	2.00
	TRIALS	-	-														
130	TEMP WHITE MASS PUMP TRIALS	3.00	100%	3.00	3.00	3.00	3.00	3.00	3.00	3.00	3.00	3.00	3.00	3.00	3.00	3.00	3.00
	MAIN INSTALLATION CONTRACT	-	-														
135	COMPLETE F/E DES/ISSUE ENQ.	1.00	100%	1.00	1.00	1.00	1.00	1.00	1.00	1.00	1.00	1.00	1.00	1.00	1.00	1.00	1.00
145	EVALUATE BIDS & SELECT	1.00	100%	1.00	1.00	1.00	1.00	1.00	1.00	1.00	1.00	1.00	1.00	1.00	1.00	1.00	1.00
150	PLACE CONTRACT	1.00	100%	1.00	1.00	1.00	1.00	1.00	1.00	1.00	1.00	1.00	1.00	1.00	1.00	1.00	1.00
155	DETAIL DESIGN & PROCUREMT	10.00	64%	5.50	5.40	7.30	6.20	8.50	8.10	8.50	8.10	8.80	8.60	9.70	9.70	9.90	9.40
160	CONST/TEST & PRE-COMM	50.00	61%	5.50	4.00	10.00	7.50	22.00	20.00	22.00	20.00	25.00	23.00	28.00	28.00	35.50	30.50
195	COMMISSION	5.00	-														
	OVERALL PROJECT	100.0	74.9	37.7	36.9	48.0	43.6	53.6	63.0	65.6	62.8	66.9	66.6	72.7	71.3	80.4	74.9

Fig. 13. Progress tabulation sheet

Project reporting and documentation

Financial reporting

For reasonable control of any capital expenditure plan it is necessary to have the financial status of the plan immediately available. This awareness can be achieved by close monitoring and reporting at both project and Capex annual operating plan levels.

Information concerning the financial status of individual projects will come from two sources – actual costs committed and expenditure arising from orders placed and payments made would be provided by the company accounting system, and the status of the project budget with planned, actual to date and forecast cost figures for each project element being supplied by the project manager as part of the project reporting procedure.

Monitoring of the annual operating plan is against the aggregate of both 'authorized' and 'unauthorized' sectors of the plan.

The status of the authorized sector is the summation of the status of individual project budgets highlighting, the authorized Capex sum, latest forecast of expenditure by financial years, costs committed and actual expenditure to date.

The unauthorized projects contained in the plan would contribute latest forecasts for expenditure planned for this financial year and future years.

The format of all financial reports would need to segregate authorized and unauthorized projects and list against each identified project the following:

☐   project number and title
☐   authorization status
☐   authorized or Capex plan value
☐   latest forecast of project costs
☐   latest forecasts by financial year
☐   costs committed
☐   costs paid to date.

Project reports

Project progress reports

As a requirement for communicating the status of projects, each project undertaken needs to provide a formal written report on progress each month.

The style and content of progress reports will be dependent on the value and nature of the particular project. The following categorizes projects as minor and major and suggests a break point at a project value of £250 000.

Minor project progress report (Project value up to £250 000)   The report will be in the form of a single page executive summary and will comprise:

- [ ] project title, project number and month of issue
- [ ] project forecasts – latest forecasts for project completion and total project cost listed against the 'as authorized' date and value (a sample project executive summary sheet is contained in 'Documentation', below)
- [ ] project status – planned and actual achieved percentage figures for overall project progress, cost committed and costs incurred
- [ ] contingency report – contingency value authorized and contingency used to date
- [ ] overview – written comment on progress, expenditure; forecasts and key issues for the current month
- [ ] critical items/variances to critical path – written statement of problems that are likely to cause delay to project completion or budget overspend
- [ ] milestones achieved – record of milestones met since last report
- [ ] next month's objectives – list of planned objectives for the coming month
- [ ] signatures – project manager signature and issue date.

Back-up documentation to support the content of the report will need to be held in the project file. As a minimum these documents should include

- [ ] project bar chart – updated with 'time now' line
- [ ] budget build-up – master control estimate format
- [ ] cost commitment/expenditure plan
- [ ] progress plan or progress curve
- [ ] progress measurement calculations and weightings.

Major project progress reports (project value in excess of £250 000)   The report will be in the form of a multi-page document and will comprise

- [ ] executive summary
- [ ] content as described for minor project
- [ ] project brief outline of project scope, objectives and key dates
- [ ] project bar chart updated with 'time now' line
- [ ] project progress curve
- [ ] cost review sheet values against each budget element for authorized sum, latest forecast, costs committed and cost paid to date
- [ ] cost plan or cost curve.

Back-up documentation will need to include

- [ ] contractors report for each contract or major work package
- [ ] procurement schedules
- [ ] schedule of scope changes/variations
- [ ] progress measurement weightings/calculations
- [ ] budget build-up (master control estimate).

Contractors' reports

Allied to internal project reporting is the requirement for regular progress reports to be prepared and issued by contractors. The following is intended for use as a guide to the level of contractor reporting and will be related to contract values and work packaging arrangements.

Listed below is the typical index for a contractor's report for a major contract (in excess of £250 000). Reports for contracts of lesser value will be derived from this format by the project manager to suit overall project reporting requirements.

Contractor's report index

- ☐ Project overview – series of concise statements recording general status of projects against various phases of project execution.
- ☐ Areas of concern – single line statement on issues that impact on project cost or programme.
- ☐ Progress review/summary – progress curves for overall project, design and construction phases; progress measurement calculations.
- ☐ Bar chart update – bar chart to employ 'time now' line to register progress for each activity.
- ☐ Financial report – report to record current status relative to original contract sums and adjustment caused by variations.
- ☐ Cash flow summary – report on anticipated turnover of contract value on month by month basis.
- ☐ Procurement schedule – summary of contractors procurement activities relating to major purchases.
- ☐ Design report – schedules identifying status of design, drawings, and project related documents.

The requirements for reporting will be identified in enquiry documentation and agreed levels contained within the contract specification.

Project close-out report

The purpose of this report is to provide a formal record of the project for future reference. The project manager is responsible for the preparation of the report and for the collation of the data relevant to his project.

The scope and nature of close-out reporting will be dependent on the value and complexity of a particular project. Set out below is a format consisting of a series of desired headings, not all of which necessarily apply to any given project.

Contents

Executive summary   General statement of project achievements relative to technical objectives, programme and costs (a sample form contained in 'Documentation', see Fig. 17)

Project scope   Definition of project brief, objectives, location, description, budget, timetable and implementation plan

Technical review   Description of implementation to include, project management, design, manufacture, installation and commissioning.

Procurement review   Defines commercial strategy. Identify major vendors and contractors employed and comment on their selection, resources, performance and financial/contractual expertise.

Programme review   Defines programming and reporting procedures, commenting on project progress against programme relating to contractor performance where appropriate

Financial review   Identifies project budget, budget allocation, actual costs incurred, under/over expenditure, variations, scope changes, contingency management

Observations and conclusions   Reviews project achievements relative to project brief in terms of technical, time and cost objectives. Describes particular problems or benefits for future reference. Identifies difficulties and successes with implementation methodology.

Health and safety file   Incorporated, by reference or appendix, the health and safety file is compiled during the life of the project to satisfy the requirements of the Construction (Design and Management) Regulations.

Appendices   Append as appropriate relevant project documents, for example, programme, progress/cost reviews, takeover and acceptance certificates.

Documentation

Sample document formats are illustrated in Figs 14–17.

FINANCIAL REPORT
CAPEX ANNUAL OPERATING PLAN

Issue No. :
Date :

CR No.	Project Title	Auth'd Capex	Latest F'cast	Prev. Years Actual	This Year Actual	Future Years F'cast	This Year F'cast	This Year AOP
AUTHORISED Projects > £250k.								
Projects 0 – £250k.								
AUTHORISED PROJECTS TOTAL								
UNAUTHORISED Projects > £250k.								
Projects 0 – £250k.								
Other Projects								
UNAUTHOR'D PROJECTS TOTAL								
CAPEX TOTALS								

Fig. 14. Financial report – Capex annual operating plan

PROJECT PROGRESS REPORT
EXECUTIVE SUMMARY

PROJECT TITLE :
CR No. :   **DATE:** (MONTH/YEAR)

PROJECT STATUS. Overall Project Progress (%) Costs Committed (£) Costs Paid to Date	Planned	Achieved
FORECASTS. Project Completion (Date) Total Project Cost (£) Spend Previous Year (£) Spend This Year (£) Spend Next Year (£)	**As Authorised**	**Latest Forecasts**

CONTINGENCY.   Used to Date (£)
Authorised (£)
Forecast Use of Contingency (£)

OVERVIEW.

CRITICAL ITEMS/VARIANCES TO CRITICAL PATH.

MAJOR MILESTONES ACHIEVED.

NEXT MONTH'S OBJECTIVES.

....   Date ....

Project Manager   **Signature**

Fig. 15. Project progress report executive summary

| | | | **COST SUMMARY SHEET** (expressed in £k) | | | | | | |

PROJECT TITLE:
CR No.:
Issue: (MONTH/YEAR)

Description	Auth. Budget	Revised Budget	Cont'g Alloc'd	Cost Comm'd	Cost Inc'd	Prev. F'cast	Latest F'cast	Vari'ce
ON-SITE								
Plant & Equipment								
Building & Civil Works								
Mechanical Installation								
Electrical Installation								
Instrumentation & Control								
Enabling Works								
Capital Spares								
Commissioning/Training								
ON SITE TOTALS								
OFF-SITE								
Project Management								
Pre-authorisation Costs								
OFF-SITE TOTALS								
PROJECT COST								
CONTINGENCY								
TOTAL PROJECT COST								

Fig. 16. Cost summary sheet

PROJECT CLOSE OUT REPORT		
EXECUTIVE SUMMARY		
PROJECT TITLE : **CR No. :**		**DATE:** (MONTH/YEAR)
PROJECT STATUS. Overall Project Completion (Date) Project Cost (£)	**Planned**	**Achieved**
CONTINGENCY. Authorised (£) Used (£) Available (£)		
OVERVIEW/TECHNICAL DETAILS.		
CONTRACTORS'/VENDORS' REVIEW.		
TRAINING.		
SAFETY.		

Fig. 17. Project close-out report executive summary

6. Accounting for projects

J. Horne

Why can accounting requirements and project requirements differ?

It was once said that America and England were countries separated by a common language. The same can be said about project managers and accountants. Many of the words and phrases used by both are the same but can have very different meanings.

Financial accounting requirements are strongly governed by the requirements of the Accounting Standards Body (perhaps this is why accountants seem so inflexible). Because of these requirements any of the requirements of accountants may, to the project managers, seem to be irrelevant to the planning, implementation and management of projects.

Project managers, on the other hand, are more concerned with tracking their spend against authorities, managing individual contracts and projecting their spend to the end of their projects. What may come as a surprise is that many of these requirements go hand in hand and it is simply one group's lack of understanding of the others' requirements that can lead to confusion or disagreement of what each requires.

Definition of words and phrases

To try to help unravel the controversy of what is meant by accounting concepts in project accounting, detailed below are the key words and phrases and their project and accounting translations.

Cost of work done

The accountants view of cost of work done (COWD) is that work which is completed and invoiced or certified at a given point in time. This will not always be the value of the work done to date, the reason being that the given point in time may have passed. If, for example, the last invoice/certificate raised was at 13 March and we are trying to ascertain the project cost at 31 March then we will need to accrue for the balance.

Accrual

From the example above we can see that there is a need to estimate the value of work done between 13 March and 31 March. There are many means to arrive at this figure from an accurate evaluation of the amount by the quantity surveyor on site to a rough and ready estimate by the project manager. Although the accountant would prefer the former in reality, the figure provided would be nearer to the latter. The only time the accountant must have accurate figures is at year end when the statutory accounts are being prepared.

Work in progress

This is the term used by accountants to explain the current level of cost on a project. In essence it is the sum of the COWD and the accrual.

From the example above:

COWD (certified at 13 March)	£1 000 000
Accrual (for work done 13–31 March)	£20 000
Work in progress	£1 020 000

Source data

Source data are just that. They are data taken from the original source. These could be in the form of invoices, applications or valuations to name a few. The data can be either internally or externally prepared.

For example, an invoice is prepared by the supplier of goods or services but the preparation of a certificate can be prepared in-house (although the original application may well have been produced externally by either the supplier or a quantity surveyor acting for either party).

An accrual is generally prepared in-house although as with the application detailed above the measurement of the value can also be carried out by a quantity surveyoy acting for either the supplier or the project manager.

Most of the data seen by the accountant are in the form of invoices or certificates. What is the difference between the two?

The two major differences for the accountant are the tax implications and the way they are produced.

Invoices

As stated above the main source for the invoices is from the supplier of the goods or services. Law requires that certain information is presented on the face of the invoice, although we will not go into the detail here, suffice it to be said that if the company is VAT registered then it is a requirement that VAT is charged on the invoice where the goods or services are vatable. The invoice will also state the tax point date. This is the date when the company producing the invoice created a liability to pass the tax to Customs & Excise. The date the invoice was raised will also be stated on the invoice. Generally, this will be the same as the tax point but it should be noted that this is not always the case.

Once the accountant receives the invoice, the tax is accounted for from the date shown as the tax point and can be reclaimed once it has been processed (even if the invoice has not been paid).

Certificates

Generally, before a certificate is produced the supplier or a third party acting on behalf of the supplier will send an application for payment. This document is not a legally binding contract for payment but is used as the basis for agreement of the certified amount (although quite often they can be the same figure). The key difference, however, between the invoice and the certificate, as far as the accountant is concerned, is the treatment of VAT. When a certificate is raised, an authenticated receipt needs to be produced by the supplier.

The authenticated receipt procedure allows a supplier (subcontractor) to issue an authenticated receipt for payment instead of a normal VAT invoice provided it shows all the necessary details required to be shown on VAT invoices and that no tax or similar document is issued. The procedure is operated by the main contractor preparing a receipt for supplies received and forwarding it to the subcontractor with the stage payment. The reciept is not valid for tax purposes until it has been authenticated by the subcontractor. The main contractor can claim input tax relief in the VAT period in which the subcontractor receives the stage payment, without waiting for an authenticated receipt although he will later have to prove to Customs & Excise that his VAT return was correct by obtaining and keeping the authenticated receipt from the subcontractor.

The benefit of operating this system to the subcontractor is obviously that the VAT does not become payable to Customs and Excise until he has received the payment from the main contractor, therefore helping cashflow.

Preparation of project accounts and financial accounts

This is the starting point of the monthly reporting on projects with the project managers. They are solely responsible for confirming valuations on the cost of work done by internal or external quantity surveyors. Because accountants place a cut-off period on valuations which does not necessarily fit in with the work periods of project managers, some of the quantity surveyor estimates may be precise measurements or best estimates.

Whichever it is, the project manager must agree the valuations (ideally at contract level) and establish the cost of work done at that given point.

This information is then passed to the accounts department, either manually or via a computerized project management system.

The accountant will compare the cost of work done provided against actual invoices that have been posted against the project. The difference between the two is accrued so that the accountant's work in progress agrees with the project manager's work in progress, that is,

COWD per project manager	£525 000
Invoices posted in accounts	£420 000
Accrual posted to accounts to agree	£105 000

The accountant will always make sure that the sum of all of the projects (after they have posted the accruals) agrees with the cost of work in done in total held by the project manager.

Accruals may either be posted into the accounts at contract level or at project level. Certainly at interim (half way through the year) or at the end of the financial year the accountant will be checking each accrual at contract level for reasonableness.

Projects within projects

There are occasions when, for simplicity or for contracting purposes, project managers will bundle projects together under one overall project reference. The budgets may then be added together to give an overall tendering budget.

For accounting purposes and for management control it may be necessary to show and account for each project separately. If this is the case the accountant and the project manager should work together at the contracting stage to ensure that all the necessary information can be separately reported and captured. Between them they will also need to decide the basis for the apportionment of management and supervision costs.

To some project managers this may seem pedantic and unnecessary; however, apart from the necessity of cost control against budget, the accountant will also be interested in the breakdown of the costs for tax purposes. By having a full and detailed analysis of spend the accountant will be in a position to optimize his tax position and minimize the risk of wrongful declaration and overpayment of tax.

Although a project manager may group these projects together to allow a full and complete review and to ensure that costs are being managed effectively, the accountant may choose to report costs at individual project level.

This allows comparisons of costs against budget or authority to be made and the right decisions to be made to mitigate identified project areas and overspend areas.

Completed projects

When projects are physically completed the accountant is able to transfer the asset from work in progress to what is called fixed assets. It is at this stage that the asset can be registered in the fixed asset register and depreciated (written down) in the accounts; it becomes an independent asset in its own right.

It is not, however, as simple as transferring a cost, in work in progress, once physical completion is reached. As both the project manager and accountant are aware there are further costs which could be incurred before the completion certificate is issued. These costs may be items such as snagging, ongoing project management costs to reach completion, claims, etc.

Therefore, before the accountant transfers the work in progress to fixed assets, the project manager and accountant should discuss potential costs which may arise, for example:

COWD per project manager	£500 000
Invoices in accounts	£400 000
Accrual on COWD raised	£100 000
Unsettled claim	£50 000
Snagging costs	£5000
Project Management costs to complete	£2000
Total accrual required	£157 000

This accrual should be reflected in the accountant's and the project manager's ledgers. Therefore the total value of the asset is

COWD in accounts	£400 000
Accrual	£157 000
Total value of asset	£557 000

This in itself is not the final stage in the management of the project. Although the asset has been capitalized in the accounts, the accountant will be carrying an accrual for the trailing charges (the costs which have been agreed above but have not yet materialized). The accountant will monitor this accrual every month, and will watch the value of the accrual decrease as invoices are charged against it. Once the completion certificate is signed off, the accountant will review any accrual held and decide whether it is now prudent to release the unused accrual. (This happens when the estimate of the accrual when the trailing charges were decided is on the optimistic side.)

Once the final review has taken place, this ends the project cycle, and it is at this point that a post-completion review of the project can take place involving the project team, the finance team and any others as appropriate, and from that will come a report detailing the lessons, both good and bad, that have been learnt from the whole exercise. These lessons should be fed back into the cycle to ensure that problems are mitigated in future projects and good practice is adopted.

7. Engineering and project department organization

S.H. Wearne

Engineering and project departments which undertake many projects need clear priorities, direct leadership, simple procedures and attention to the longer-term development of staff. The projects exceptional in their size, risks, urgency, novelty or special importance to a customer should be run separately.*

Tasks

Organization is a means of enabling people to achieve more together than they could alone. Or it should be. How and how much to organize a department should therefore depend on how and how many people need to work together. How far to define a structure of jobs should thus be decided on the basis of the work to be done and how people are dependent on each other to do it. The cases described earlier and reports of other examples show that the main tasks of engineering and projects departments are to:

☐ prepare project proposals, to the extent required to assess their feasibility, identify risks, estimate cost within a stated accuracy, and plan the remaining work needed to complete projects
☐ provide advice on ideas and problems
☐ propose new projects of potential value to the organization
☐ design, specify and detail projects, to the extent required for ordering manufacture, construction, completion, testing and acceptance by the customer. Typically this work uses half of a department's total man-hours
☐ develop the expertise required for future projects; prepare project budgets and programmes
☐ prepare departmental budgets and programmes

---

* The word 'customer' is used here to mean the person or organization which is purchasing a department's services, whether by external contract or internal order. The word 'project' is used to mean any package of work, whether it is for an external contract or an internal order.

☐ monitor the achievement of quality, cost and time targets
☐ review the results of all the above and recommend improvements.

The above is not a standard definition of what a department should be doing. It is a checklist, for use when planning and reviewing a department's work. It is a reminder of what may or may not be required.

Organizational needs

The lesson of all experience is that every project needs someone responsible for it from start to finish, though this may change from person to person because of individual workloads or suitability.

Organization

The organizational problem is that most of the projects in hand in a department are not large enough for people to be dedicated to only one project at a time. Attention to each project may also be interrupted by the needs of other projects or by delays from a customer or other parties. Small projects may need some or all of the time of perhaps only one person, plus some services from others. Many projects may follow a similar pattern and call on much the same expertise, but regular relationships are interrupted. The organization therefore needs a system for setting priorities to employ people effectively, rather than a formal structure which separates them into groups or project teams.

Always there needs to be the technical expertise among the staff demanded by the nature of the projects, innovations, the problems and the risks, and experience of applying this expertise to the type of customer and the range of work. An engineering department therefore needs its experts, but with the flexibility and the ability to work on more than one project at once. An engineer may have several roles – leader of one or more projects, expert in a subject, and adviser on a customer or type of system.

A standard system of organization that suits all demands has not yet been evolved. Probably it never will be, because companies differ in their objectives, culture, work and circumstances. The following two examples illustrate the structure of engineering departments responsible for many projects. The two examples are not intended to show how all such departments should be organized. They show instances of choices of structure to provide the basis for going on to discuss the potential advantages and disadvantages of alternative ways of organizing people to achieve their tasks.

Structures – a customer organization

Projects

Company A is a manufacturer. The engineering department discussed here provides a service to one of the company's three factories. The factory is thus its main customer in an internal market. Most of the projects in hand by the department for the factory are for extensions and improvements to operating plant, site services and methods of maintenance. This range of projects is expected to continue to be the main work but with increasing demands in them for greater system reliability, cost-effectiveness, automated systems to reduce man-hours, and engineering choices that meet environmental and other 'green' criteria.

The project work consists mainly of the specification, planning and checking of the procurement, installation and handover of new or replacement mechanical and electrical systems, equipment and software, and at least minor civil engineering, demolition and building work. Innovations by suppliers of equipment are utilized, particularly to increase reliability and automation, but the main decisions in the design of a project are in adapting engineering experience to the conditions of a site.

A large project such as to build a new production line would be the responsibility of a separate project team formed from corporate headquarters, hired temporary staff and some members of this department but operating largely autonomously.

Formal structure

The department's formal structure is shown in Fig. 1.

Under the chief engineer* in this example the primary division of work and responsibilities is by type of work, in three principal groups of staff:

☐ Technical services group – responsible for services to the factory works engineers and to projects, and divided into three specialist sections.

☐ Engineering services group – responsible for the design and specification of projects, in three sections of specialist engineers. It also acts as consultant to the works engineers.

☐ Projects group – responsible for planning, estimating, the quality assurance system and monitoring all project work. The construction and installation section is responsible for the design and specification of water, waste, building and civil engineering work.

---

* Titles and their significance vary from company to company. Typical titles and roles are shown in the examples in this chapter.

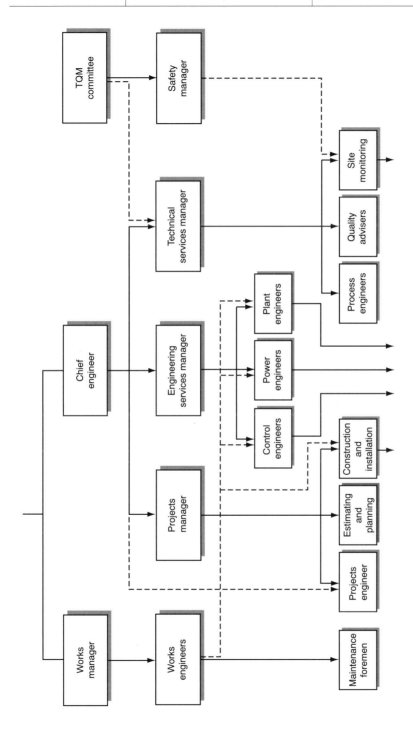

Fig. 1. Factory engineering department – company A

In each group the next division into sections is also mainly by type of work. The sections are labelled by what the people in them are qualified to do by training and experience.

An engineer allocated the work of design of a project is responsible for its delivery and costs. Thus the responsibility for each project is added to the permanent structure.

The one person designated 'projects engineer' is responsible for risk analysis, delivery and costs of the projects which depend upon work by more than one engineering group.

Contract terms and the choice of a contractor are the joint responsibility of the works manager, the chief engineer and the company purchasing officer (not shown in Fig. 1), with the decisions on the smaller ones being delegated to the purchasing officer, works engineers, and the department's engineer on the project. Contractors' work on site is supervised by the works engineers. Control of changes, reporting and certification of payment due to a contractor are joint responsibilities between them and the engineer in the department who specified the work.

The projects engineer recently joined the department after being project manager of a large project. He has the continuing task of monitoring that project's manufacturing performance, and also the new role in the department of planning the use of the resources, assessing the department's effectiveness and the consequences of customers' changes to projects and priorities.

The chief engineer, group managers and project engineer meet monthly to review the future needs of the department. The group managers are not involved in the detail of projects. In practice they act as consultants to the groups on unexpected problems.

Project decisions

The organizational problem of the department is that it has a fixed number of staff but much of the work comes at short notice, in three ways:

☐ From works engineers – their requests amount to about 60% of the total work for the department. Some are for factory improvements and extensions which have to be planned in advance and the main work for them can be foreseen. More of it is for factory problems which are rarely predicted and usually urgent. The works managers hold the budget for all this work, and therefore they decide their priorities.

☐ From technical services – they propose projects to improve processes or site conditions. The chief engineer has a budget for preparing such proposals, but the works manager has the budget to make the improvements and decisions to proceed tend to come at short notice to suit production and when cash is available.

☐ From corporate level – requests to study or to provide information for studies of possible large projects for site. These amount to 10% of the total work. The demand for these studies is sometimes known in advance, but this work is not welcomed as most of it is returned for unexplained revision.

Until recently all requests for work by the department were supposed to come in through the chief engineer, so that he could plan the best use of the resources. This was impracticable, as it took up most of his time and tended to delay the many small requests from the works engineers. All levels in the department considered that this service was important for their recognition in the company. In a recent change the specialist engineers were therefore authorized individually to commit 50% of their time to that service, directly to works engineers or through the technical services group, and the projects engineer became responsible for advising on the priorities for the rest of the department's work.

The dashed lines in Fig. 1 indicate these relationships needed in addition to the formal system of management:

- [ ] Works engineers to the construction, control, power and plant engineers, as mentioned above.
- [ ] Chief engineer to the projects engineer – for directions and reports on the projects dependent on more than one engineering group.
- [ ] Total quality committee – setting the standards for engineering and process quality and site monitoring. The chief engineer, works manager and safety manager are members of this committee.

Corporate services

In the past 10 years the company has invested in very few large projects, compared with its growth in 15 years up to 1960. Decentralization of choice of projects to each factory has therefore been the policy, as described above, with retention of only a few central planning and technical staff, on the basis that staff could be hired temporarily for large projects when needed.

More larger projects are now expected, some on a new site in the UK but at least one at any time at this factory. Associated companies abroad also are requesting design and project services. The company is therefore developing central staff specializing in project economics and innovations in manufacturing systems. Development of a central project engineering group could follow. It might be expected to become a corporate service with specialization in automation and reliability. From this, financial and technical standards might be decided centrally and applied to all the company's projects which could conflict with the present independence of the factories as separate profit centres.

Each factory now chooses its contractors and contract terms. They mainly employ local contractors, but when employing national contractors a factory could be in dispute with a contractor which is favoured by another of the company's factories. The three factories' purchasing officers now meet regularly to avoid this. With the growth of larger projects more larger contracts are likely, and with them the employment of more national contractors. Some centralization of contract strategy might follow, but with the risk of removing the accountability of a factory for all the decisions on managing its projects.

Centralization of decisions and amalgamation of staff can achieve some economy of scale, but may result in loss of motivation, slower response to factory problems, and a lack of feedback from the local consequences of remote decisions. In the present system the direct links between works engineers and individuals in engineering services achieve detailed accountability for the results of projects and feedback to the next.

Structures – a contractor organization

Projects

Company B sells equipment and systems by contract to the orders of other companies who use them in their factories to handle materials and products. Its customers vary in their size and frequency of orders, so form a varied and varying external market.

Nearly every order is a new project requiring the design and supply of a system and equipment to meet a customer's specification of the performance and layout required to suit use of them. Usually only a few machines are sold in one order, but parts and sub-assemblies can be repeated from previous projects. The company therefore undertakes the design of many projects largely requiring adaptation.

The company designs and tests prototypes to try and test innovations ahead of orders, to add to the standard range of equipment offered for sale. Detail can be varied for each order and some orders to design are accepted, but as competition causes short delivery times to be offered the policy of the company is only those variations essential to obtaining an order should be accepted.

Subassemblies, components and control software are bought from subcontractors. The company installs and tests them, but may require the specialist subcontractors to provide advisory supervision as well as written instructions.

The company thus sets out to supply machine modules and subsystems as already designed and proved. Success in this depends upon delivery times, predicting the demand and achieving reliable innovations.

The company is a subsidiary of major holding company. It trades with its sister subsidiaries as a customer or contractor, but is independent as a business and in all its project and engineering decisions.

Formal structure

The company's formal structure is shown in Fig. 2.

Under the general manager the primary grouping is based upon stage of involvement in a bid and then an order.

The marketing and sales department is expected to be largely self-sufficient in marketing and in bidding. The systems engineers are in this department to plan design of systems to meet customers' requirements. They report their recommendations to the chief engineer, but he and the engineering groups under him are involved only if he requires it for exceptional design features or the systems engineers request his advice.

When an order is received the chief engineer appoints one of the project engineers to be responsible for the detailed design and preparing specifications for the subcontracts, producing the information needed for erection, testing and commissioning, and for co-ordinating all other internal work for the order. The project engineers have the direct services of a design office. The other engineering staff are specialists providing services to all the projects.

Development is initiated by the chief engineer, in principle independently of customers' current enquiries and orders, but in practice consists of a mixture of specialists for enquiries and longer-term ideas from within the company. This extends to a budget for development work by the systems engineers under the direction of the chief engineer.

Project decisions

One problem is limiting the cost of designing for bids which may not lead to orders. Ideally, sufficient detail would be decided before bidding to be able to commence placing subcontracts immediately an order was received from a customer. The man-hours this would cost before the certainty of an order to pay for them is considered to be too great. The principle adopted is that only unusual risks and innovations required should be studied by engineers before an order. The result is that the project engineers have to decide much detail very quickly after receiving orders. Their technical and co-ordinating role is crucial to successful completion of orders. All other staff therefore then have to operate as a service to the project engineers.

A problem within this is the handover between the systems and the project engineers. The latter argue that they should be asked to comment on bid proposals before they are final, and also that a proportion of the systems staff time should allocated to handing over when a bid leads to an order. The solution being tried is that both groups have agreed on a checklist, as shown in Fig. 3, of information on the customer, his objectives, contact names, style, reasons for the terms of the bid, risks and possible changes which should be compiled before a bid is made so as to provide a

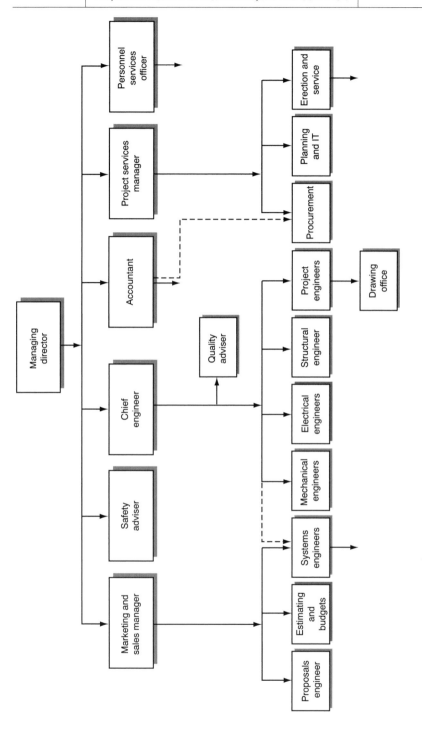

Fig. 2. Contractor's engineering department – company B

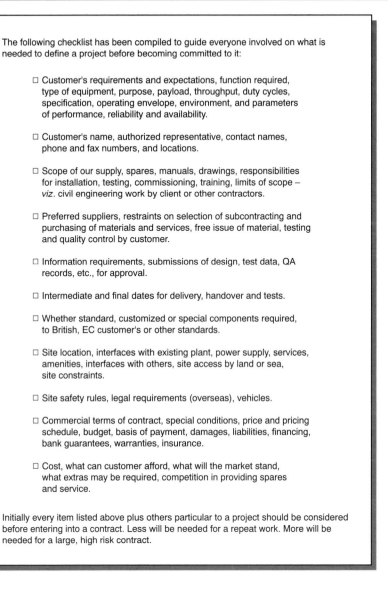

The following checklist has been compiled to guide everyone involved on what is needed to define a project before becoming committed to it:

- ☐ Customer's requirements and expectations, function required, type of equipment, purpose, payload, throughput, duty cycles, specification, operating envelope, environment, and parameters of performance, reliability and availability.

- ☐ Customer's name, authorized representative, contact names, phone and fax numbers, and locations.

- ☐ Scope of our supply, spares, manuals, drawings, responsibilities for installation, testing, commissioning, training, limits of scope – *viz.* civil engineering work by client or other contractors.

- ☐ Preferred suppliers, restraints on selection of subcontracting and purchasing of materials and services, free issue of material, testing and quality control by customer.

- ☐ Information requirements, submissions of design, test data, QA records, etc., for approval.

- ☐ Intermediate and final dates for delivery, handover and tests.

- ☐ Whether standard, customized or special components required, to British, EC customer's or other standards.

- ☐ Site location, interfaces with existing plant, power supply, services, amenities, interfaces with others, site access by land or sea, site constraints.

- ☐ Site safety rules, legal requirements (overseas), vehicles.

- ☐ Commercial terms of contract, special conditions, price and pricing schedule, budget, basis of payment, damages, liabilities, financing, bank guarantees, warranties, insurance.

- ☐ Cost, what can customer afford, what will the market stand, what extras may be required, competition in providing spares and service.

Initially every item listed above plus others particular to a project should be considered before entering into a contract. Less will be needed for a repeat work. More will be needed for a large, high risk contract.

Fig. 3. Checklist for project requirements

formal basis for the handover. Independently of this the senior managers have also agreed that some staff should henceforth be moved around these roles for their broader development, but also with the expectation that it will improve communications between all sections.

This system depends upon the chief engineer's experience in foreseeing problems and initiating innovations. Neither he nor any one person automatically receives regular feedback of the consequences of the systems and equipment supplied to customers. Major problems are likely to be referred back from the section trying to solve them and complete the detail, but the structure does not otherwise provide a means of accumulating experience of the consequences of decisions.

The dashed lines in Fig. 2 show the link for development work between the chief engineer and the systems engineers. Also shown is the authority of the accountant for agreeing all procurement orders are within budget before commitment to subcontractors.

Choices in principle

Conflicting objectives

The examples show that organizations have to try to meet two potentially conflicting objectives:

- ☐ Providing people and other resources for the sequence of activities essential to the transitory demands of each project.
- ☐ Achieving economy in resources by using them as continuously as possible.

Structural choices

Various ways are possible for grouping people in departments and sections in large departments to try to meet the above two objectives. The stages of the work provide one possible basis, as in the last example described.

Another basis for organization also illustrated in most cases is the expertise needed in analysing problems so as to group people according to their specialist knowledge and to draw on this when deemed necessary for making decisions.

There are examples of all of these to be found in practice. Each way of dividing the work in a company has some merits, but a consequence of any system of specialization is that it is used as the basis for locating people.

The tendency to group specialists together is natural when they share the use of data or services, but their resulting separation away from other groups greatly reduces communications between them. Any separation of people dependent on each other in their work can hinder the co-operation needed to meet objectives. This combined with geographical separation can prevent the exchanges of information needed to solve novel problems in the abstract and uncertain processes of design. Specialization is valuable, but with it there must be attention to the needs of each project and the flexibility to innovate.

Location

Two principles thus apply to minimize problems of communications and co-operation:

- [ ] As far as possible locate together all who are dependent on information from each other.
- [ ] Individuals and groups should be located close to where problems may arise.

Project teams

The obvious means of achieving attention to each project and its objectives is to dedicate people to a project from start to finish. The cases illustrate limitations on doing this for small projects and when it is uncertain whether a proposal will be followed by an order to carry it out. The principle remains an important one.

The larger the project, the more it may be both important and practicable to do so. Putting together all the people and resources required for a project directly under a manager responsible only for meeting the objectives of that project to form what is known as a project task force should concentrate attention and help motivate all to meet the customer's needs. Choice of this system is clearly logical as a means of concentrating on clients' and customers' requirements, as projects are the productive work and all other activities have value only as a contribution to a project. It puts together the people dependent upon information and decisions, and should therefore be coherent in reacting to demands and to new problems. Designating a separate project team involving all affected can also be a means of achieving an organizational transition, utilizing people's interest in the technical novelties or other features of a project to draw them into a changed system. It is the system that most tests the project managers, as each for his project has the range of responsibilities of a general manager. This can be a useful part of a company's scheme for developing general managers.

On the other hand the needs of a project for various skills, etc., are only temporary and may fluctuate. Skills and other resources are not shared between projects, which can be wasteful at the time and fail to accumulate experience for future projects. Sharing of these needs between projects rather than having self-sufficient project groups may therefore be much more economical. Experience can then also be shared and reserves of resources drawn upon when required. Dedication to one project requires people to work themselves out of a job by completing their stage of the work. A customer may demand the formation of a dedicated team, because of his concern about a supplier's arrangements to complete a commitment. The cost to the supplier is obvious, and therefore may have to be imposed.

Grouping entirely by project is a principle that is appropriate for a 'one-time undertaking ... that is infrequent or unfamiliar ... complex ... and critical', as concluded in an early study of some examples in companies in the USA. Except in these conditions it is more logical to give preference to a system for sharing resources, adding to this a secondary means of linking the decisions for each project.

Grouping by subject

In this system people can be grouped according to the main branches of technology: civil engineering, metallurgy, etc., and within these in groups specializing in subsidiary branches, such as soil mechanics, structures, highway design and others within civil engineering. Specialisms can also evolve which draw together parts of several branches for application, control engineering being an example.

An illustration of the detailed choices possible in following this principle for the division of work is given in Fig. 4.* It is a polar diagram of the choices possible within a civil engineering consultant's staff working on the design of bridges, drainage and road details for highway systems. The vertical variable in Fig. 4 is the extent to which people in this section could specialize on bridges, or drainage or road detail. The horizontal variable is the extent to which they specialize in design studies, analysis, safety, drawing or preparing contract specifications. With the greatest divisions of work a member of the section could be specializing in the analysis of bridge problems. With the least, all members of the section could be accepting any part of the work. A mixture might be best, particularly to be flexible, but in any choice there would also be a need to integrate the work for a project. Such a diagram is a means of analysing the resources and utilization of these in a group, one potential use of this being to indicate the recruiting or training needed.

Specializing in a subject and becoming expert in its application is the way in which 'know-how' can be accumulated, the lessons learned from mistakes and the experience of similar problems shared. All such specialisms are learned by study and application. Further study and meetings with corresponding specialists in other companies to exchange experience in that branch of technology are necessary for individuals to keep up to date in their subject. The company advances its expertise by investment in research, by individuals keeping up to date, and by recruiting.

Members of subject groups can act as reserves for each other and train new members so as to form a pool of expert resources. Grouping by subject is therefore the logical principle for establishing expertise for application amongst projects when required and for accumulating experience from one

* This diagram is derived from a proposal by J.S. Hulshoff Pol, a Netherlands' management consultant.

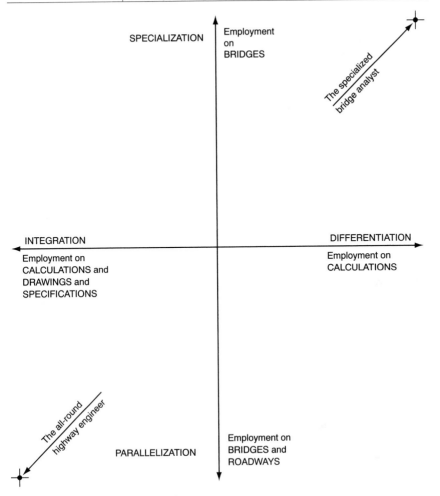

Fig. 4. Specialization within a specialization

project to the next. It is suitable for linking design with development work in the company or by outside specialists.

Specialism by subject is vital in engineering to enable individuals to master and to advance sections of the expertise needed in the design of projects. It is therefore a feature of most organizations. This system offers careers in specialization. It makes less demands on section managers if their task consists of the types of problem familiar to them which arise in their section's specialist work. If they are also required to resolve conflicts between project and specialist objectives the demands on them are much greater.

The disadvantage of this system is that work divided amongst specialist groups has to be fitted together to meet project objectives. There can be confusion because objectives vary from project to project, and there can be conflicts of interests because the responsibility for results is indirect. In the solution of a design problem a specialist may wish to spend more resources to reduce the uncertainties in his predictions of the consequences of choices. Other people on the project may consider that a decision must be made and some risks accepted because of limitations on cost and time. At its extreme this is a conflict between the safety of the fully analysed solution and the reality of economy and consequent risk. Such conflicts tend to arise when a project is divided amongst a large numbers of specialists. Any one group can regulate only its own part of the work. Their specialism may be their dominant interest. To them project objectives may seem uncertain and can become secondary in influence on their decisions.

These consequences of separating people into subject groups may be acceptable if the division of knowledge of a project as a whole is required to achieve secrecy or security. Coherence and common knowledge and understanding between groups is more usually required to achieve economy and collaboration in solving problems.

Grouping by stage of projects

In the alternative of dividing work in series, groups of people can specialize in a stage of decisions with an immediate objective in the evolution of each project. One example described such a 'vertical' division in the relationships between companies. It is found in many companies. In such systems experience is accumulated in each stage. People with similar prior training who enter different departments become experts in solving problems in different stages of work. Their output is the starting point for another group.

When innovation is sought in the selection of new projects the initial stage of studies of ideas could be separated for this purpose, so freeing one group from the detail and the urgency usually required in the subsequent design of the projects selected. This first stage can be subdivided, so that one group is free to consider innovations and another group to specialize in the evaluation of proposals. This system can be appropriate when the company is engaged on innovative and on adaptive projects, to separate the initiation of these two types of proposals. It can be valuable when the demand for one of these classes of work fluctuates, as only a few people need specialize in it and others from outside or from other groups can be allocated to assist them when necessary.

Division of the work by stage can be applied to the remainder of the design sequence, as in the example shown in Fig. 2 and discussed earlier. After the selection of a project the process can become the responsibility of a group specializing in the main stage of decisions. A further group could inherit the results of their decisions and be the specialists in detailing and preparing the other instructions for manufacture, etc., and the subsequent stages including information on how to use and maintain the product.

In this system the first group has the objective of assessing the choice of projects, the second the solution of design problems and the third the issuing of the results. From assessment of whether a project should be chosen the work changes to deciding how best it should be realized and then to specifying what needs to be done to make it. By division of the work into these three stages each group has an immediate objective. The system reflects how the nature of problems and decisions change as the work proceeds.

A disadvantage of this system is that it depends upon linking from stage to stage. The acceptance of prior decisions may not be successful because of failures in one stage to foresee problems that arise in a later stage or because of human uncertainty about accepting the risks of inheriting the consequences of decisions made by others. A system of organization based upon the principle of division of work into stages is therefore most likely to be effective where projects require only adaptive design and when the system has become familiar to many of its members. In such a system there should be regular cycling of people forward to earlier stages of the work so that experience of consequent detail is renewed in the groups making the first decisions in design. In some cases this is possible because people can also move with a project through the stages to work with others specializing in a stage. Some apparently uneconomic employment of people may be needed if the continuation and the timing of projects depend upon uncertain changes in demand.

Stage by stage grouping forms a natural basis for the control of costs and progress. Such a system depends upon the capacities of the groups being planned to meet the flow of work from one to the next. Most of the problems of the group managers are repetitive, arising in the linking of the stages.

Functional grouping

The term functional organization is used in textbooks to mean in effect grouping by subject and by stage combined. The term is avoided here so as keep clear the two principles, and to avoid confusion with the idea of 'functional' supervision by a set of experts.

Levels of decisions

Levels of authority are a common basis for decisions in organizations. On this principle the authority to make decisions can be handed down a hierarchy as the work proceeds. Many examples illustrate this, the usual principle being to delegate the more detailed and less risky decisions. The sequence of decisions to be made in the design of projects is a logical basis for defining authority to make these decisions and it would probably be widely understood in practice. This is seen in the examples in companies. Stages of decisions by levels is found at the initial stage of design in companies and within specialist groups where the group leader takes part in analysing and solving the initial problems in the group's work for a project but delegates the consequent detail down one or more levels. These divisions by level are most often found combined with divisions of the work into stages.

The disadvantages are that it is inflexible in utilizing the abilities of individuals and that it does not establish the links which are needed between related choices for solving parallel problems in the sequence of decisions. As with the associated principle of division by stage, division by authority level is likely to be appropriate only in settled conditions of adaptive design.

Project co-ordination and control

At the start of this chapter we stated that one lesson of all experience is that every project needs someone to be responsible for it from start to finish. For small projects this can be part of the role of the engineer responsible for defining or detailing a project, as in the two examples. For larger projects the same could apply, but if nearly a full-time task it becomes a separate role. In the example shown in Fig. 1 the projects engineer has this role for larger projects.

The need varies according to the number of people employed on a project, the interdependence of their work, the extent of unexpected problems, whether they are located together, and whether most are familiar with these conditions. The simplest arrangement is to add the co-ordinating role to a person already employed on work for the project, to keep communications simple. The effect of this may be limited by the role being that of co-ordinating others, rather than being their boss in the formal managerial system.

If the size of a project makes its co-ordination a full-time task, a separate person can be given the role on behalf of the manager controlling the resources essential to that project, in a 'Staff' position as indicated in Fig. 5. The important principle in this is that the staff position is not an additional level in the hierarchy.

The projects engineer shown in Fig. 1 is in effect in this role in relation to the chief engineer (so are the safety manager in that organization and the safety and quality advisers in Fig. 2).

In theory, people in staff roles act only as an extension of the line* managers' capacity. They have to be accepted as givers of decisions authorized by the line managers.

* 'Line' meaning with authority over resources.

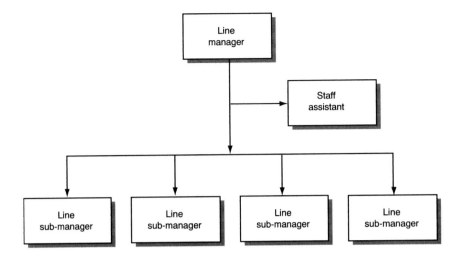

Fig. 5. Line and staff roles in a management hierarchy

One or more people can be in this position. The arrangement is similar in principle if other departments provide a service to a line manager.

Like anything logical, the line-and-staff arrangement works when all concerned understand it. In times of conflict or uncertainty it may be limited in its effectiveness because decisions remain dependent upon the line members of the hierarchy of authority.

Matrix systems

Companies and public authorities have evolved what are called 'matrix' systems of management to achieve leadership of a set of projects which are sharing the resources of groups of specialists. An instance of this is to be seen in the relationship between project engineers and the other engineers in Fig. 2. As indicated in Fig. 6, matrix systems can be more complex and organic, particularly in the extent of the formal authority of the project leaders relative to that of the group managers.

The project and the specialist leaders should theoretically all influence decisions in a matrix system. A project leader may be responsible for quantitative decisions affecting the cost and programme of his project, whereas the specialist managers may be responsible for qualitative standards in allocating people and other resources to each project. If so, the specialist managers have to act as consultants to members of their groups once allocated to a project. A matrix system thus provides opportunities to employ leaders with different skills and knowledge in these two types of managerial roles.

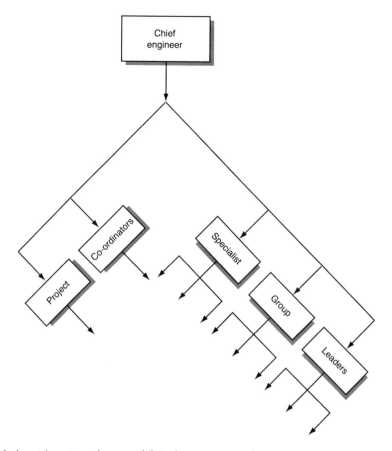

Fig. 6. A matrix system of responsibilities for resources and projects

The matrix principle is awkward to try to depict in the two-dimensional diagrams used to show systems of organization. The usual convention in drawing organizational diagrams is that the line of authority is vertical. Collaboration is thus horizontal. In a matrix these are compounded. Thus the example in Fig. 6 is drawn on the skew.

Matrix management or internal contracts?

Observations of matrix systems indicate that there can be problems in them between project managers and the specialist group leaders concerning the allocation of resources to a project and the quality, cost and timing of the work to be done by them. The group leaders should earlier have agreed on specifications, budgets and programmes for every project, but may have done so some time before a project starts and then only in sufficient detail to get a budget or a contract to proceed. This may not ensure that adequate resources are available when a project calls for them.

One means of avoiding most or all such problems is to treat each specialist group's work for a project as in effect a contractual commitment. If it was to be purchased by contract from another company there would normally be a prior process of investigating the potential supplier's capacity and understanding of the work required, followed by an invitation to offer to do it for a price and specified quality and delivery. Procedures for progress reporting, inspection, changes and resolving problems would also be agreed before they might be needed. The same are in effect needed within organizations, not through legally enforceable documents but by written or oral definitions of what is expected of others rather than assuming that these are understood and agreed.

Final comments

Employing people in departments which specialize by function is the tradition in many industries. With this the departments and some individuals within them specialize in a stage of work for projects. This can be efficient provided the work is repetitive and there is systematic feedback to show where problems need attention. Job rotation to develop individuals can also help to achieve good communications and project feedback.

The alternative of dedication to a project team is appropriate to concentrate on a novel or otherwise special project, or to achieve security. Sharing in a matrix system is more common in order try to meet the shorter-term project needs and achieve economic use of longer-term expertise.

No one structure is likely to the best for ever. The proportion of unforeseen and uncertain problems of projects in all industries has increased because of greater uncertainties in predicting markets, technological changes, public controls and the continuing growth in the complexity and the economic risks of new projects. A project can less and less be defined once and for all at the start and the subsequent decisions delegated down among groups divided under a hierarchy of managers. More problems require continued linking at levels in the hierarchy which do not have the capacity for this unless they are augmented.

In all but the very smallest companies the working relationships between people should therefore be reviewed regularly. Otherwise individuals are likely to concentrate on their own work and its importance as they see it. Specialists are particularly likely to do so, because they are motivated by their interest in their work. And if treated as advisers rather than as members of project teams they may also tend to concentrate on quality rather than delivery times and budgets. Specialization in technical and other tasks is valuable in enabling groups and individuals to accumulate expertise and concentrate on different parts of the whole. The risk is that no one of them can then know or makes the time to find out how well their relationships are operating.

For engineering managers this means that their responsibilities for what may be the answers to project problems and technical questions must be extended to attention to *how* the decisions on these are made.

Further reading

Rutter P.A. and Martin A.S. (eds) (1990). *Management of Design Offices*. Thomas Telford, London.

Smith N.J. (1995). *Engineering Project Management*. Chapters 8 and 9. Blackwell, Oxford.

Wearne S.H. (1993). *Principles of Engineering Organization*, 2nd edn. Thomas Telford, London.

8. The management of projects

D. Major

Safety

The bottom line here is if somebody asked you to lay down your life, or spend the rest of your life in a wheelchair so that the project could be completed on time, to budget and to quality you would think he was out of his mind. But if you are managing a project and you do not set up adequate safety systems and standard procedures you are asking somebody to do just that.

Project health and safety plan

Principal contractors are now required by law, under the Construction (Design and Management) Regulations to maintain a project health and safety plan (it is originated by the planning supervisor) though many competent companies and contractors have been doing this for some considerable time now. Health and Safety Executive (HSE) guides provide a guideline as to how this document should be produced, and if liberal doses of common sense are applied this can be a very useful aid to the success of a project. The project health and safety plan sets out an overview of the systems and procedures to be followed on a project which should make it reasonably safe and help mitigate health and safety risks, but it is only part of the armoury of documentation that helps to create a reasonably safe project. It is important to understand here that there are health and safety risks associated with everything we do, and in planning a project, the overall objective is to reduce the risk, not necessarily to eliminate the hazard, as this may not be possible or reasonably practical in all instances, and so we are talking about risk management here.

Planning for safe design and construction

One of the most serious deficiencies in this area in relation to health and safety risks on a project is the failure on the part of most designers, usually through lack of relevant experience, to design structures and machinery which are safe to fabricate and erect, safe in use and safe to maintain. This stresses the need right from the start of the project, to involve the construction manager, and operations and maintenance personnel. Their expertise and knowledge in terms of the buildability, operability and maintainability of a project is vital in providing guidance to the designers to ensure that what they have designed may be safely built, operated and maintained. It also stresses again the need to plan the commissioning of the project at the start, so that operations and maintenance issues may be addressed. Obviously as the detailed design progresses, answers to many operational and maintenance design issues will become available. One of the major milestones on most process engineering projects where control systems are provided is the production of a functional design specification for the control system. This document gives a vital insight into how the plant will be operated and hence upon its completion, the detailed design is usually advanced enough to determine how the plant may be maintained by the carrying out of a 'Hazop' study. At or shortly after that stage of the project a detailed awareness of the operational and maintenance risks associated with the plant, and equipment may be obtained. If this stage is reached early enough it can provide vital guidance to the designers to avoid expensive design changes later on. What is certain is that if these resultant design changes manifest themselves at a tertiary stage of the project, they will be even more expensive to implement. The mandatory production of design risk assessments in accordance with the Construction (Design and Management) Regulations definitely assists this thought process and if done properly will help avoid expensive mistakes.

Method statements and risk assessments

Method statements and risk assessment are now required by law under the Construction (Design and Management) Regulations. However, these are vital aids to the overall planning of a project and should be done at the earliest opportunity, albeit that they will probably be in draft form and have to be rewritten perhaps more than once at a late stage. They can provide a detailed understanding of work to be carried out, and consequently make cost and time planning more accurate at an earlier stage. If properly produced and worked to, they certainly reduce health and safety risks to personnel working on the project, simply because they force the individuals concerned to plan out properly what they do before they do it.

Creating the appropriate attitude towards safety

Site safety and/or project safety has traditionally been viewed by those in the construction business as a necessary evil. It is rarely viewed as a cost cutting exercise, which is essentially what it is. Producing method statements, risk assessments, and health and safety procedures forces people and companies to think through and plan what they are going to do before they do it. This generally means that they do it more efficiently and more quickly, and, therefore, more cheaply. Equally, if and when accidents occur, their direct cost usually pales into insignificance when compared to the indirect cost of delay, disruption, and in the case of serious injury or fatality the lowering of morale and the consequential productivity losses. Put simply, managing a project in a safe way saves money. It could also save lives.

Quality

Cost and time are very important, but failing to achieve adequate quality can be catastrophic as can the pursuit of a level of quality which is far in excess of that required. If Rover built Minis like its parent company builds 7-series BMWs, its engineers would be very happy but its accountants would be heartbroken. The same logic applies to other engineering projects. Usually at the onset of a project the price and the time are fixed, but the scope and quality have yet to be fully defined and these are the result of a meeting of minds between client, main contractor, subcontractor and vendor. It is important to bear in mind that there is no such thing as a free lunch. In short, quality costs.

Project quality plan

Project quality plans are a fairly recent innovation, and the writer has been producing similar documents for over 20 years, variously described as project implementation plans, project execution plans, and encompassing project procedure documents. These things are different names for a document which seeks to demonstrate to the reader how it is intended to do the project, what processes, procedures, and systems are to be used, and hence give an indication of how quality, cost and time are being controlled (albeit that cost and time can be inferred to be controlled by project quality plans, few of them explicitly state this). A project quality plan needs to be brief, to the point, practical and realistic. If it is too lengthy the team will not bother reading it, and if it is too complicated to follow they will not understand it, and if the procedures and systems that it outlines are unworkable, it will very quickly be abandoned. The important thing about a project quality plan is to set realistic, achievable, measurable goals that are both specific to the project and minimalist (and hence cost-effective).

Quality inspections

As mentioned above, quality inspections are best carried out by members of the project team and they draft in other expertise as and when required. The logic is simple. If the project electrical engineer does not know what he wants in a motor control centre, then how can he expect anyone else, vendor, subcontractor or client to know? So it is important that quality inspections are carried out by people who know what the performance requirements are, as well as what the specification says. The timing of quality inspections is important. They should be intended to hit milestone events in the production of products whether off-site or on-site and should be as independent of opinion as possible.

Quality standards in documentation

The project quality plan should be an all-embracing document that sets out the quality of drawings or specifications and seeks to set out systems, procedures and methods to be followed which will lead to a good level of total quality management in everything that is done on the project. Investing time and effort in checking and approving drawings and specifications is very worthwhile, and in the writer's experience is very rarely done to anything like a reasonable standard. A good acid test is to ask non-members of the project team (although for obvious reasons appropriately technically qualified people) to review the specification and/or drawing and provide constructive criticism. Although this will expend man hours and hence cost money, in the writer's experience it is cost effective in terms of the expensive errors that tend to get avoided. Also, main contractors should not be afraid to use their client's engineers as a free checking services or indeed the subcontractors. Contrary to what is sometimes thought, these people are not usually fools and quite often have more time to spend on the exercise than a main contractor's own personnel, and certainly most vendors and subcontractors will have a greater level of expert knowledge in a particular field than any members of the main contractors or client's team. It is worth seeking their opinion in a constructive fashion (if possible prior to the competitive tendering exercise, for obvious reasons).

Specifications and drawings, if produced to a high standard of quality in terms of making their content practical, achievable, realistic and in keeping with current practices can reduce the costs of purchased goods and services from vendors and subcontractors dramatically. The converse is also true. Badly written specifications can cause very expensive mistakes on projects. The writer had an unfortunate experience (among many over the years) in this respect when a stainless steel fabrication for use in a food plant was accidentally shot-blasted with iron shot rather than with stainless steel as a result of a poorly worded specification.

Planning

It has often been said that there are three important things to get right in project management – planning, planning and planning. Whereas this is not strictly true, planning is of vital importance, and by planning, I do not mean producing a bar chart, sticking it up on the wall for 3 years and marking it up with coloured pens. Planning needs to be carried out at all levels, by all individuals working on a project, and it may take many forms. One of the most important considerations is that the planning methodology, processes and procedures followed must be kept as simple as possible and as easy to understand as possible. Producing a 2000-activity network in technicolour on a personal computer and showing critical path analysis in three shades of red may impress one or two people, but will probably be incomprehensible to most and serve no useful purpose whatsoever.

Critical path analysis

There are several very good planning software packages on the market, such as *Power Project*, *Microsoft Project* and *Plantrak*. All of these are perfectly adequate for most projects, and, as a result of the advances in personal computers over the last 5 years, even a fairly substantial project can be planned in overall terms and monitored by a single planner.

An overall project plan must be produced for even relatively small projects and it is vital that this is accepted as being a live document. It is no use producing one and just marking it up when going through the project. Because of logic changes, and the fact that no-one can predict the future, it will very soon become totally useless as a management tool. However, if one of the above software packages is used sensibly, a single full-time planner with the part-time aid of the whole project team can initially plan the project from start to finish and then, with a little bit of help from the rest of the team in terms of gathering information, the programme can be updated and rescheduled every month to show the project team not only exactly where they are, but where they are going in terms of critical activities and forecast completion date. Manpower histograms, progress curves and cost curves are all 'nice-to-haves' but are 'icing on the cake'. Without knowing criticality month to month, week to week and the forecast completion date, a project manager would be 'up the creek without a paddle'.

Design

When a project falls behind it usually starts with slippage in the design work. It is important to understand that nearly all projects in the design phase can have the work split into two categories. Conceptualizing usually involves a bit of lateral thinking, kicking ideas around and generally running down one or two blind alleys before it is possible to come up with the concept design to run with. And then there is the detail design. Where the line is drawn between the two will vary from project to project and has to be a matter of common sense, but the essential difference is the former conceptual stage is a lot more difficult to predict than the latter detail stage. Therefore, when planning a project, a generous period should be allowed for the concept stage (think of a number and quadruple it) but it is necessary to be very draconian and strict about sticking to the timetable. The detail stage will be (after going through the concept stage), relatively easy to define (and hence guess activity durations for). If design starts to slip, more resources should be allocated to it, because it is generally cheaper than trying to accelerate procurement, construction, testing, or commissioning. However, in the euphoric honeymoon period of most projects, this generally gets forgotten and design is usually allowed to slip on most projects – an expensive luxury. It is vital to understand and plan for the design co-ordination function and be realistic about when design interface information will become available from vendors, subcontractors, clients and the principal contractor.

Procurement

The planning of procurement successfully is sometimes very difficult because it involves estimating periods of time taken by a variety of individuals within one's own organization and within others. However, the periods of time estimated are usually based on historical information. It is important to allow sufficient time for technical and commercial evaluation/ negotiation, and this is probably the most imponderable part of the exercise. However, it is worth noting that the length of time spent on the technical and commercial evaluation of tenders and quotations from vendors can have a profound effect on the final negotiated cost and this is one area where the quality of the exercise should not give way to pressure of time. Allow a fair period of time to do this.

Subcontracting

It is important when planning the process of awarding subcontracts (and then the subcontractor implementing them) to have a full understanding of the process and procedures that must be gone through, and what systems are going to be followed. To illustrate the point, if the standard subcontract documents generally look like 15 volumes of *War and Peace*, then it may be prudent to allow a little more time for this activity than if they have been rationalized into standard format and are readily understood by subcontracting and main contracting personnel alike.

Construction/installation

As stated previously, it is vital at the earliest stage of a project, and certainly prior to the completion of the conceptual design, to produce an overall plan for the project, (usually broken down to level 3 planning) and as a part of the exercise, the expertise and knowledge of the site construction manager should be used to plan the job, in particular to take account of logistics, site access, and buildability issues. Preparation of draft method statements at this stage helps a great deal with this exercise, and such issues as site layout, temporary works, site compound, laydown area, cabin set up and assembly areas should all be planned in some detail at this stage.

Testing

It is important at the early stages of planning to decide what off-site and on-site testing is to take place and plan for this work. Do we have to use a mobile testing laboratory or any heavy duty equipment for which space has to be allocated?

Commissioning

In the haste to get whatever it is built, quite often the planning of commissioning does not take place until the tertiary stage of the project. This can be a very expensive mistake, because sometimes when it is thought that the project is on programme, or even ahead of programme, there is the sudden realization that there is not enough time to commission it. It is vital, right at the start, prior to the completion of concept design, that commissioning is planned out in some detail, including what on-site and off-site sampling and testing is going to take place, the planning of training and the planning of the preparation of operations and maintenance manuals.

Financial planning

The most appropriate phrase that springs to mind under this heading is 'let's not kid ourselves'. Projects are sometimes authorized and/or budgets approved because somebody is deliberately underestimating the cost in order to achieve the authorization or approval – whether an employee, a company, the client organization, the contractor or the subcontractor. However, if there is serious interest in determining what the actual cost will be, it is important to bear in mind that whatever the contracting strategy for a project is, it usually involves any one or a combination of the following: using directly employed or contract people working for the organization, employing vendors, subcontractors and main contractors. The cost plan for the project should be kept as simple as possible in simple spreadsheet format. Whether this is the client organization, the main contractor, the subcontractor or the vendor, all that it is necessary to know is the original planned cost against each item, the cost to date, the forecast cost to completion, and breakdown of the variations which have arisen or are predicted to arise. The moderately intelligent clerk can put together historical information such as the value of purchase orders and subcontracts and come up with a set of figures. The clever bit, usually masterminded by the project manager and his commercial manager (the commercial manager usually being a cost engineer or a quantity surveyor by profession) is to accurately guess the rest. Let us be clear on this, these are just educated guesses because nobody can foretell the future. Bearing this in mind the estimation of cost for envisaged variations against subcontracts and vendor purchase orders is a skill developed by the above persons based upon many years of relevant and appropriate experience. If the individuals concerned do not have the relevant and appropriate experience the exercise will be a failure and the guess is likely to be wildly inaccurate. The cost plan for any project should be kept simple and cost centres should reflect the contracting strategy. In a typical capital project or projects this will be broken down into the form of an in house man-hour estimate designed to predict and monitor the cost of the internal man-hours spend by the project team (usually presented as a simple bar chart over the calendar period of the project) and a spreadsheet showing the planned costs per day per week or whatever. The rest of the report is a spreadsheet containing planned actual and forecast costs as mentioned above for each cost centre and this is cross referenced to breakdown sheets which list the variation costs, both actual to date and forecast, giving a brief explanation of what they are about and whether the cost figure is actual or estimated. If the project cost report contains any more information than this, it simply confuses the readers and distracts them away from the important issues to be addressed during the project review meeting. On most projects a project review meeting should take place once a month and cost reports should be updated along with planning and other reports to produce a written project report by a particular cut-off date in the month. Because financial people, accountants and surveyors love figures, these basic rules, which are all geared towards keeping it simple and therefore easily understood, are rarely followed and some project financial reports turn out to be so voluminous and complex that they are virtually incomprehensible and so the principal objective of producing them, which is to examine the financial health of the project, is lost. As a consequence of this, during project review meetings the fundamental issue of what cost movements have occurred in the reporting period and what new cost movements are envisaged, rarely get properly debated. This tends to have very serious consequences on the project, although it may be in the project managers best short-term political interest, and more fundamentally the organization involved misses the opportunity of an early debate as to what went wrong, why it went wrong, and what lessons can be learned for the future. If this debate does not take place the mistakes are likely to be repeated, again and again.

If the project financial report is kept simple, then the time taken to update it each month is shorter. This minimizes the amount of time the project manager and commercial manager have to spend on the exercise and not only results in a direct cost saving in terms of their man hour expenditure but more importantly releases their valuable time to be spent on more important issues. Conversely, if the financial reports are long and complex and the project commercial manager is spending a week per month updating and preparing them, then serious commercial difficulties can arise on a project because he is not available to deal with them.

On the subject of forecasting costs, when an original cost plan is prepared for a project, the knowledge and experience of the project manager and commercial manager is applied to the 'guesstimation' of cost movements and hence the 'guesstimation' of the total cost of the project. A particular cost centre may, for instance, be the supply and installation of the centrifuges and based on experience elsewhere the individuals concerned may choose to put, say, a 10% contingency sum against this cost centre and so the principle is repeated throughout the cost plan to come up with a final forecast cost for the project. The important thing about this exercise is that it must be realistic even if it is intended to produce a totally spurious report for the rest of the world. It is vital to know what it is really thought the costs are going to be and it is only by going through the cost plan line by line and asking the basic question 'What is likely to go wrong?' and giving an honest answer that the result is likely to be a financial forecast that is realistic. Realistic means an overestimate of what the project will cost. It will never be possible to guess it to the correct number for any project and it is better to have a budget that is too big rather than one that is too small.

There are many estimating software packages on the market. Like all computer hardware and software systems these will not do anything that cannot be done by oneself, and equally in many text books on the subject of cost estimating projects there are rules of thumb relating to factors and percentages. All of these can be useful in the appropriate situation but there is no substitute for common sense and honesty.

Management of off-site work

One of the most common myths of modern-day project management is that if a vendor or subcontractor is tied up with a suitably hard contract, then the risk of delivering on time to quality and cost is somehow lessened and therefore it is unnecessary to go to the same lengths to control the production of the project to cost, time and quality. This is obviously nonsense and many a project has failed due to failure to understand this.

Planning

The planning of the off-site work cannot be left to the vendor and/or subcontractor in totality, although they must produce and work to their own programme. This must be rolled up into the main contractor's overall programme and reported on in the same manner as on-site work. Failure to plan and monitor off-site work will ruin a project just as effectively as any on-site problem.

Expediting

Another commonly held myth is that the vendor or subcontractor will somehow take care of this themselves. It is not practical or economically viable to expedite every off-site manufactured item and component. However, an early strategic review of the expediting needs (what is critical, what is on a long-lead time) is vital to the successful planning and implementation of a project. Expediting and quality audits must be carried out at sensible intervals, with realistic objectives being set at realistic times so that the expediter can take an educated guess as to whether the equipment is going to arrive on time, as well as whether it will meet the quality standards. In the writer's experience, expediting is best carried out by members of the project team (not subcontractors and not in-house full-time expediters). The appropriate members of the project team know what equipment is required, when it is required and are generally the best qualified to carry out the expediting and quality audits. They should be able to draft in expert knowledge as and when required in areas where they are not proficient.

Quality audits/inspections

As stated above, quality audits are best carried out by members of the project team who should know the specification for the piece of equipment, and also know the functional requirement for the piece of equipment, so that intelligent, practical and common sense decisions can be made (preferably on the spot) when a non-conformance situation arises. Nothing in the writer's experience sours the relationship between main contractor, subcontractor or main contractor and vendor quite like the vendor or subcontractor being told that he must do something simply because it says so in the specification, without being given an explanation of the real reason why it must be done (as quite often there is not a very good reason – it is simply that the specification is incorrect or the auditor does not know what he is talking about).

The wise auditor always remembers that the subcontractor or vendor has been chosen partly because they are experts in their chosen fields. If they do not comply with the specification, there is a small chance that they might actually know what they are doing, and there may be a very good reason for not conforming. Be warned, however, slavishly following specifications can create a lot of grief. Be prepared to question whether the specification is correct in the circumstances.

Design (subcontractors and vendors)

Depending upon the nature of the work package, the subcontractor's or vendor's design work could be extensive, and it may be appropriate to monitor it in much the same way as the main contractor's design work might be monitored. Where an extensive amount of subcontractor's or vendor's design work is being carried out, in the writer's experience, it is always wise to base the appropriate engineer from the main contractor's team in the subcontractor's or vendor's premises. This speeds up decision making and the flow of information, but it can lead to commercial difficulties.

Variations and claims

Whether one is referring to off-site or on-site work, the key to success here is to identify as early as possible any additional cost items, be it the direct cost of the items, or the cost of prolongation (preliminaries etc.). The reaction of most engineers to cost is that along with time it is somebody else's problem. Generally, engineers working on a project believe that their primary objective is to obtain engineering excellence, irrespective of cost and time, whereas what is usually required is engineering adequacy, on time and on budget. One of the most effective ways of dealing with this problem is not to leave it to the engineers, but to introduce a system of paperwork whereby anything that looks vaguely like an increase in cost or time, gets flagged up to the project manager/commercial manager and/or quantity surveyor so that it may be dealt with in a timely and proactive fashion. It is perhaps worth mentioning that quantity surveyors usually favour leaving everything until the end of the subcontractor's or vendor's contract period and then trying to pick the bones out of it.

This is akin to trying to write the minutes of the meeting 6 months after the meeting occurred. Do this, and the project will fail. It is important that a proactive culture is created in the project team, and in particular in regard to any events which have a cost or programme implication. One of the major benefits in dealing with claims and variations as and when they arise is that clouds of uncertainty do not hang over the head of the main contractor, subcontractors or vendors, taking their minds off the job in hand. Failure to deal with variations and claims early enough can cost a project dearly. If a client, main contractor, subcontractor or vendor is sufficiently upset by the failure then their co-operation and effort can be withdrawn (sometimes so subtly that it is difficult to prove).

Management of on-site work

The best approach to this problem is to imagine setting up a business in production engineering, because, in fact, similar management issues and problems are faced.

Site logistics and planning

In preparing the overall programme for the project, it is best to mark up a site plan with lay-down areas, compound areas, lock-up area, positions for site cabins and plan through the movement of materials including double handling where it may occur.

There is no substitute for a project (of any size) having a full-time on-site planning engineer, who will primarily look after the long-term view (forecast date to completion, criticality). With a planner working in close liaison with a site manager and site engineers, buyers and design engineers, the short-term view (three week look ahead weekly programme reviewed weekly on site) and the long-term programme can be established, monitored and controlled with ease. There are many aspects of site logistics and planning that require attention and if any one of these is not right it can cause severe damage to the success of a project. However, it is important to ensure first and foremost that the project team has the right quantity and quality of people (always overestimate – it is better and cheaper to have people with spare time on their hands than people who are fire-fighting all the time), and the right site equipment and accommodation. By carrying out the forward planning exercise and writing early method statements, eleventh-hour panics and inability to allocate a space to something can be avoided. This may sound like common sense, but it is amazing how many expensive and not budgeted for temporary access roads, car parks and lay-down areas have had to be established on construction sites through lack of forward planning. With forward planning it is sometimes possible to combine temporary and permanent use and the temporary needs of construction can sometimes influence the design of the permanent works.

Mobilization and site establishment

Most construction sites for medium to large projects should take no more that four to six weeks to fully mobilize and for this to happen successfully, what is required and where it will be sited needs to be thought through and planned meticulously. This should be done by the site manager and the project planner in consultation and collaboration with engineers and buyers. The site staff accommodation should be generously sized and planned as one would plan an office in a production plant, with adequate furnishings, heating, lighting, ventilation, information technology equipment, photocopying facilities, and mess facilities. Failure to plan this properly and implement it can lead to severe disruption on a project and low morale which will lead to lack of productivity. A tea lady is a relatively low cost – engineers, planners and site managers make expensive tea.

It is generally far more cost effective for the main contractor having planned the whole project from start to finish to actually provide the subcontractor's site staff accommodation in the form of Portakabins to be shared with the main contractors and the client's staff. The subcontractor will have to provide his own security stores and other items of plant hire. But it is important to recognize the benefits of forcing subcontractor, main contractor and client project teams to share accommodation. This cuts down barriers and forces them to work with each other, and streamlines communication.

Building relationships, communication and meetings

As mentioned above, building the right working relationships between the parties is vital to the success of the project. All parties should strive for openness, honesty, good effective communication, and to streamline the communication process to ensure that the right people know the right things at the right time and it is important here to remember that old adage, 'It is not what you say, it is how you say it'. The benefits of this approach may not be immediately obvious, and because of conflicting priorities subcontractors, clients, main contractors and vendors may feel they have a vested interest in keeping each other in the dark from time to time. Obviously some matters have to be confidential, but this form of overt deviousness is not in their best interests financially on a project. Keeping each other in the dark and lying to each other creates bad feeling and conflict, leading to both people and organizations becoming defensive and wasting time and effort justifying actions that are past and therefore not concentrating their minds on the action they must take in the future for the overall good of the project.

If an individual is particularly disruptive, then he or she, whether employed by the subcontractor, the main contractor, the client or the vendor organization, should be removed. Poor working relationships are the root cause of many a failed project.

Project culture

Project culture is like company culture, only it is project specific and unique to the project. It's not the same as the client company's culture, or the main contractor's culture or the culture of the individual vendors and subcontractors, but it is rather the result of all these cultures being mixed up on the project. It is possible to create a successful project culture, even if some of the vendors and subcontractors have bad company cultures. However, the culture of the main contractor and the client organizations must be good or the project will fail, simply because even though the project team might strive to create the right project culture, this will be in direct conflict with the parent company or companies, which usually have to provide some form of support to the project team, and hence the project will be likely to fail. This has often been put another way: project management cannot be successful without general management support. If the company is managed badly, whether it is the client or the main contractor, or both, then the project is very likely to be managed badly.

Clients should ensure that the main contractor is thoroughly vetted, its premises are visited, the personnel are introduced and references are taken up. Contractors should do the same, being careful with whom they do business, because clients can put them out of business.

Relationships between client, contractor, subcontractor and vendor

As mentioned above, creating the right working relationships is vital. With mutual respect, mutual fair treatment, openness, honesty, and an understanding that client organizations, main contractors, subcontractors and vendors will sometimes have differing objectives, but do have a right to pursue those objectives, a project team can work wonders and achieve a very high level of success. It is vital that the main contractor and its team take time out to get to know the individuals with whom they will be dealing and in particular the opposite numbers in the client organization, the subcontractors and the major vendors. It is important that money is invested in social events in this area, both in terms of collective nights out (even if only for a pizza, or going bowling) and business lunches. These help break down barriers, and enable people to chat socially, instead of simply talking about work all the time. Generally, the better people get to know each other, the more difficult it is to have misunderstandings and serious conflicts, and it is also important to understand that should a serious conflict arise, if individuals in opposing organizations have good personal relationships, then they can often reach an accommodation and a solution to a problem which is less extreme than it might otherwise have been, because they are more willing and able to acknowledge the other's problems.

Conflicting objectives

To illustrate this point with an embarrassing level of crudity, the client generally wants the project completed yesterday, for nothing. The main contractor wants to be paid the earth for the project, take forever to complete it, and get away with the minimum standard of quality. The main contractor also wants to preferably drive his subcontractors and vendors into bankruptcy by getting them to do their work for as little money (preferably for free) as possible, and the subcontractors feel the same way about their vendors, their subcontractors and suppliers. Maybe that's not really the way it is, but it is important to understand, particularly for the main contractor's project manager or the client's project manager, that there are fundamental conflicts in the objectives of the different organizations involved. However, all of their businesses benefit ultimately from successful projects, not from unsuccessful ones, and fundamentally, most of them, and most of the individuals working for them, want to succeed because people generally prefer success to failure, whatever form it takes. Generally, the reason why people, companies and projects fail is through misunderstandings and poor communication.

Moral standards

If we could all stop lying to each other, there would be more successful projects in the world. Sounds simplistic, does it not? But there is a lot of truth in it. The client lies to himself and to the contractor by tying the contractor into a price and delivery and/or standard of quality that he is fairly confident the contractor cannot meet, and then pats himself on the back thinking, 'Oh, I've done a good deal here'. The contractor commits himself to a programme and cost for which he is fairly sure he cannot do the job, but congratulates himself on winning the job in competitive tender. And subcontractors and vendors do much the same sorts of things, both with their clients and with their servants. In the real world, it is often necessary to lie internally and externally to win a project, particularly in competitive tendering. However, what most organizations, main contractors in particular and clients to a lesser extent, are very slow to wake up to is recognition that this process goes on, and when the main contract is awarded and/or the capital expenditure is authorized, that these are 'head on the block' occasions when it is appropriate to take stock of the situation. There have been many instances where a Capex authorization has been made by a client organization to carry out a project allowing less money than it needs to do the job and/or time, and equally many occasions where main contractors have been awarded a contract that initially they felt they could not do in the time or for the cost.

The usual reaction to this is to go on lying internally and externally, producing monthly project reports containing lots of pretty charts which do not really tell anybody anything, rather than facing up to the problem. If the problem is faced early enough, it is possible (the writer has on several occasions had first hand experience of this) to reassess the situation, face up to it, and re-engineer or replan the way out of it. It should not be forgotten that from the client's perspective, the usual reason why a project is taking place is to meet a business need, and through the normal processes that client organizations go through, by the time a main contractor has been awarded a contract, one is usually looking at a well over-engineered scheme and not enough time left after a protracted tendering exercise to complete the project. If the project scope, cost and timing are reviewed properly at the start of the project, and a little lateral thinking applied, with the co-operation of all concerned, it is often a relatively straightforward matter to pull the 'fat out of the fire'.

Similarly, during the implementation of a project, if openness and honesty are practised and preached by the whole of the project team, then people are more likely to volunteer good ideas, better ways of doing the job, quicker ways of doing it or providing a better quality design. This can benefit the whole project, and, if the same moral standards are applied to the sharing of financial benefits associated with design changes or programme changes, then the individual people concerned, whether they work for client, contractor, subcontractor or vendor, will be better motivated to come up with good ideas, and the project will benefit. High moral standards are important for a project manager and his team, just as they are for a board of directors in terms of the general management of a business.

Maintaining morale

Morale is a very vital ingredient in the success of a project. It is important to realize that with most engineering projects, one is dealing with a huge, diverse cross-section of society, from the degree-qualified software designer, to the man who digs holes in the ground with a shovel and every level in between. It is important for a project manager to have an understanding of what motivates these people, what makes them happy, and what upsets them, and to make a conscious effort to pander to their psychological needs in this respect. The important issues here are – take the right steps to team build both off-site and on-site, whether this concerns people digging holes in the road, or a scientist doing a concept design study in an office. It is necessary to ensure that they have a good standard of accommodation, properly heated, lit, air-conditioned, good standard of furniture, appropriate information technology equipment, with respect to site work adequate changing and drying facilities, adequate transportation arrangements, and that adequate arrangements are made (tea lady, canteen, etc.) to ensure that they are able to eat and drink at work without wasting a lot of time. As well as taking care of their physical needs, it is important that appropriately designed documentation is made available to them to demonstrate the chain of command in the project team structure and who needs to know what, when and how. Included in this arrangement must be clear, easily understood routine arrangements for communication between the project team and those physically carrying out the work, including provision for regular meetings, both formal and informal, tool box talks, training lectures, etc.

All of the above, if structured and conducted in a practical common-sense way, will help maintain a high level of moral through avoiding frustration.

Team structure

The team structure will be dictated by the nature of a particular project, and its programme. It must contain sufficient expertise to cover all of the technical areas covered by the project, and sufficient quantity of personnel to cope with the volume of work involved. It is vital that each and every member of the team is issued with a document usually entitled 'Project procedure document' or 'Project implementation plan' or 'Project quality plan' or a combination of all three, that describes in very simple terms what the team structure is, who is responsible for what, who communicates with whom regarding what, the project objectives, the project scope, the project programme, the project cost plan and who liaises with whom with regard to the client organization, the vendors and the subcontractors. The document or documents must be brief and to the point, and they must get the message over. Confusion can be at best a time waster, at worst a cause of disasters, and the cost implications of misunderstandings can be horrendous.

Motivation and incentives

The primary motivation for most employees is their pay cheque. Ideally they must believe that they are slightly over-paid – that is to say, they are not tempted to look around to increase their salary. Losing key personnel part way through a project can be particularly catastrophic and as project management generally moves a lot quicker than general management, the effects of losing an individual can be quite profound. It is therefore advantageous to try to retain personnel, at least until the project is completed. The payment of project specific incentives can be particularly useful but even more important is the creation of the right project culture and working environment, as it tends to be the little things over a long period of time that can demotivate people very thoroughly, such as the inability to get a decent cup of tea at work, or having to risk life and limb in order to obtain access to their workplace, or finding that their computer has been taken by one of their work colleagues because there are not enough computers in the office. The greatest motivators and incentives are really making work as enjoyable an experience and rewarding in the financial sense as possible. For much of the time it is the little things (such as having a tea lady instead of a coffee machine) that can make all the difference to employees.

Setting and achieving objectives

I normally hate to use jargon but SMART objectives are the best – Specific, Measurable, Achievable, Realistic and Timed. If a truly open culture on a project has been achieved, and communication is as good as it can be, then the setting of objectives is easy because people are not afraid to tell the truth, no matter how harsh it is, and that way, provided that there are effective communications, people do not set and/or take on impossible objectives. The challenge on a project is not to see how fast something can be done, it is to set aside a realistic time period for the carrying out of activities. If things are being rushed then sometimes the result is substandard quality and/or safety, which can increase cost. Planners and quantity surveyors should be particularly aware, whether dealing with individuals or companies, to only set SMART objectives, be they financial or time-based.

Creating a 'can-do' mentality

I am sure that I am not the only project manager who has felt on many occasions that members of his team, the client's team, the subcontractor's team or the vendor's team are trying to find ways of not meeting the deadlines on the project. The usual pattern of events is that people fail to plan their workload properly, get behind, and rather than taking action to catch up, will think of a dozen good reasons why it is not their fault and therefore they do not have to catch up.

No one thing creates a 'can-do' mentality in a project team. It is rather a combination of most of the things that have I mentioned, particularly good team building, good relationships, high moral standards. The raising of morale can naturally lead to a 'can-do' mentality on a project and if the main contractor's and/or the client's project team demonstrate a 'can-do' mentality, this tends to be infectious and rubs off on subcontractors and vendors. Of course the converse is true also.

Project services

Time, planning, systems and procedures

The term project services comes originally from the American oil industry. It generally refers to the time estimating and cost-estimating of projects. In several industries, planners and cost engineers are often one and the same person and the financial and time planning is done from one office, usually referred to as a project services office.

Whether these exercises are done by one person, or two or more persons, and whether they are done as an integrated exercise or separately, there are some basic rules that must be followed if one is not to become unstuck. There are three basic rules that must be followed if the systems and procedures being used for the effective planning and monitoring of cost and time on capital projects are to be carried out to a reasonable degree of accuracy and provide project management with the timely information it needs to control and manipulate the cost and the timing of a project. These three rules are keep it simple, keep it simple and keep it simple.

Myths and mistakes made by people working on a project are discussed later in this chapter. However, one of the biggest problems with the planners, quantity surveyors and cost engineers is that most of them seem to have a passion for complicated documents. This tendency must be avoided. The documents must be kept simple and easy to understand. Where these people add value to the project and save money is by using their experience to improve the accuracy of their forecast and to advise project management of the potential pitfalls ahead with regard to delays and/or cost increases, and thus hopefully help avoid problems. Try to ensure that all of the members of the project team understand that time is money. Any delay to a project usually delays subcontractors and/or vendors and somewhere down the line will incur costs. Failure to appreciate this has often led to ridiculous situations on projects where, to illustrate the point, the subcontractor may be asked to extend his on-site period by 3 weeks to carry out £3000-worth of work in the third week, his on-site establishment costs (which the main contractor will have to pay) may be £10 000 per week. The project manager and project team need to be ahead of the game to see this problem coming and if cost engineers and planners are tied up producing voluminous and complex documents they are not affording themselves the thinking time to foresee this sort of problem and avoid it.

It is also worth mentioning that in carrying out cost planning and looking at the cost of variations, the financial implications of delay should not be overlooked. The cost of implementing a variation may be £5000, the delay and disruption to other subcontractors could cost you £50 000. It is important to fully understand the implications of variations before implementing them. Sometimes early warning of this situation may be given by the planner and/or the cost engineer.

Contracting strategy

Objectives political and practical

It is important right at the start of a project to write down the objectives in priority order and it is preferable to include this information in the project implementation plan, project quality plan or whatever the document is called that describes the objectives and how these will be achieved. The objectives will be split into political and practical and it is important that these are recognized and written down. If the objective is at the end of the project to have a satisfied client or satisfied client's consultant, and maintain a good working relationship with the major subcontractors and vendors, then it is worth writing this down in the implementation plan and making sure that the project team are made aware of these objectives. Failure to do this will probably result in the objectives not being achieved. At the start of the project it is necessary to prepare the document that describes the objectives and how these will be achieved. As a part of that document it will be necessary to outline the contracting strategy. Simply put, a contracting strategy is how the contract is to be split into packages of work, who is going to do the packages of work, on what basis, and how the interfaces between the packages of work will be managed. Packages of work will split into the work being carried out by the project team, the vendors and the subcontractors. This exercise has to be done in parallel with, and definitely not before, a programme is prepared for implementing the project, and the programme cannot be prepared until the scope of the project is defined. The sequence is strategy, scope the work, plan it and then resource the programme. At the stage where resources are being planned to achieve the programme it is important not to underestimate the work involved in the management of the interfaces. This will give a very strong guide as to what the contracting strategy should be. Many a project has failed because so many subcontractors and vendors were employed and the client's or main contractor's project team was not big enough or experienced enough to manage the interfaces.

Experience

Another important factor to take into account is experience, be it client, main contractor, or subcontractor. It is necessary to understand what potential subcontractors and vendors are good at. It does not matter how good the contractual wording is on a contract document, if organizations or individuals have been employed to do something that they are not capable of doing, they will fail. Which brings us on to the important subject of risk assessment in terms of contracting strategy. It is very important to ensure that companies and organizations are employed to do what they have demonstrated they can do in the past. This reduces risk because it reduces the risk of them failing to perform.

In the same way as when interviewing people to be members of the project team, for which appropriate experience and qualifications together with a track record are being sought, the same principles must be adopted when seeking to find a subcontractor or vendor to come up with a work package. To arrive at the most appropriate contracting strategy for a project, obviously cost is a primary consideration, but bearing in mind all of the above it is important not to be fooled into believing that the cost will be low by going a particular route, failing to take into account a realistic cost estimate of the interface management costs. For the most part these comprise the direct cost for the personnel required to manage the interface, the cost of variations and delay and disruption costs which will arise because they will not manage the interfaces completely efficiently and the cost of any overall effect that this may have.

Documentation

As stated above, the contracting strategy must be stated in a project implementation plan right at the start of the project so that the project team and others understand what the strategy is, what the work packages are, and ultimately the identity of the organizations involved who are going to implement the packages.

It is also important to understand that other major players in the project such as subcontractors, client organizations and major vendors also need to know the contracting strategy. They, like members of the project team, need to know not only what they are required to do, but also what other companies are required to do. It is important that the project implementation plan or project quality plan, or whatever title is given to the document, is issued to these organizations as well. It is always useful in this document to have a simple list of vendors and subcontractors, listing the name of the work package, the name of the company employed to carry out the work, the name of the point of contact at the company, and the name of the point of contact in one's own organization, together with the appropriate telephone numbers. This ensures a direct line of communication and all of the people in all of the organizations know the identity of the individual in their respective organizations through whom they should funnel information. This helps avoid misunderstandings between organizations and streamlines communication.

On the subject of the documents which comprise the enquiry documents, purchase orders and subcontracts, again the most important thing is keep them simple and therefore readily understood. Many industries and many companies, particularly in the UK, have not yet learnt this lesson but the good news associated with this is that they are keeping a lot of people in the legal profession gainfully employed. The simple acid test as to whether one has achieved the utopian ideal of having simple documentation which is readily understood as far as contract documents are concerned, is to ask a member within the organization but not directly involved with the project to read the subcontract document or vendor purchase order and indicate what the company is supposed to provide, when it is supposed to provide it, how much it will be paid to provide it and what the quality criteria are. If many organizations carried out this exercise they would be pretty horrified by the results. If the subcontractor vendor purchase order documentation is simple and easily understood, it avoids misunderstandings, claims, disputes and, ultimately, court cases. The other interesting point to bear in mind is that if it is not possible on paper to describe the safety, quality, cost, scope and time commitments to the subcontractor and vendor then it is reasonable to assume that the main contractor does not understand them. Therefore how on earth can one expect the organization employed to understand them.

Planning and managing the interfaces

This is probably one of the least understood areas of project management. It is not possible to plan the interfaces on a project and budget for them until the scope has been reasonably defined and a detailed programme prepared for the implementation of the project. From the programme, if it has been prepared correctly, one may generate a resource histogram which would determine the quality and type of individuals needed on the project team. An important element of the workload is interface management and this goes on throughout the project right up to and including handover. It takes a number of forms but the principle ones are exchanging information between client and main contractor; exchanging information between client, main contractor; and subcontractor; exchanging information between client, main contractor and vendor, exchanging information from subcontractor A to subcontractor B and exchanging information to and from main contractors, subcontractors and vendors. The quality of this exchange of information and its timing are absolutely critical and can mean the difference between project failure and project success, that is, project dramatically late and dramatically over budget or project dramatically ahead of programme and dramatically under budget. Do not underestimate the implications of getting the planning and management of interfaces wrong and understand that when employing a subcontractor and vendor, they go through a design and procurement phase themselves, and many of the questions and answers that arise during that period cannot reasonably be predicted but may have a profound effect on the work of others. To mitigate the risks associated with this, if effective and regular systems and procedures for the exchange of information are set up, managing interfaces should not do too much damage to the project. Even on a small project, the passing of the information to-and-fro through the subcontract or vendor detailed design and procurement phase and through the installation, testing and commissioning phase of the project, can be a full time job for a number of engineers. Therefore it is necessary to ensure that a realistic assessment of this is made right at the start of the project and reliance should not be placed on site supervisors to manage the questions and answers associated with design issues as this will lead to dangerous mistakes.

Personnel

Plan and resource

The most important element of any project (and indeed any business) is the personnel. Successful projects are not made by having appropriate systems and procedures in place. Successful projects are 90% made by having the right quality and quantity of people doing the right things at the right time. In the late 1990s many businesses are claiming that they understand that their business is a people business, that it is all about people. And the fashionable fads of the 1970s and 1980s computer systems, management consultants, Just-In-Time, Canban, and all the other buzzwords have finally been ousted as frauds in so far as they were proclaimed as the panacea for business ailments. In fact they are refinements of systems and procedures adopted by businesses and project teams which if one gets them right can help, but they will not do the job. Ultimately, what does the job is the people, and whether the planner uses *Artemis 2000* on a mainframe, or a pencil and a very large piece of paper, it is the quality of thought processes, the application of knowledge and experience and self-motivation that will make that person's part of the project a success or a failure. Because project management is a high-risk business, it is very easy for a £2 million building that is supposed to be built in 18 months to end up costing £4 million and taking 2 years. Equally, a £15 million process plant could very easily cost £30 million. Architects, engineers, and project managers may say that enhanced quality has been achieved in some way, shape or form, but the bottom line is that because of failure to manage the project, quite a lot of money has been wasted. With this in mind, quite obviously, the decisions made by the project team, can cost or save millions of pounds, quite literally, and all of the decisions made by most of the people on the project team will affect cost in some shape or form. It is important to properly plan resources so that people are not overloaded because of one's dependence upon the quality of the decision making. It is sensible to afford them sufficient time to make the decisions properly, rather than have them rushed off their feet working 12- or 15-hour days. Once the scope and programme has been prepared for a project the resource plan must be prepared. It is important to overestimate the number of people required to manage the project. It is far less damaging financially to have people who are managing a project underworked rather than overworked. However in the writer's experience, resourcing is very rarely done properly, it is more the norm than the exception that people working on the management of projects are overworked and therefore tend to make more mistakes than they might otherwise have made. The net effect of this is usually to increase overall cost.

Quality and quantity

The best approach by far is to employ the most experienced, competent and expensive people that can be found to manage one's project and ensure that when resourcing the programme there is a greater quantity of people managing the project than theoretically needed. Following these basic rules one will have a proactive project team ahead of the game in saving time and money. This may not necessarily be a cost-effective approach on all projects, but certainly for most projects, the cost implications of the mistakes made by people who are not sufficiently experienced and knowledgeable or by people who are overworked can be horrific.

The writer has had many experiences of such situations. In one instance an individual performing in the capacity of project engineer responsible for design co-ordination on a major multi-million pound project was inexperienced and under-qualified. The cost implications of the mistakes made by this individual in one year exceeded £100 000. His cost to the company was less than £35 000 per year and the cost of an appropriately experienced and qualified person would have been of the order of £45 000–£50 000. Hence it was not cost effective to employ the cheaper person. Whether the client's project team, the main contractor's project team, or the subcontractor's project team, the people on the project team managing the project need to be of the highest calibre available, and there needs to be sufficient of them so as not to deny them the thinking time necessary to ensure that the quality of their decisions is appropriate. That being said, the decision-making process has to be very rapid on most projects and so a compromise has to be reached.

It is also perhaps worth mentioning that as well as the right quality of personnel, one needs the appropriate experience. Assuming that the project manager is appropriately qualified in terms of experience, he is the best placed individual to determine the appropriateness of the experience of the individuals he requires on his team. It is therefore very foolish to dictate to a project manager the people that he must use. Allow him to pick his own team. To give an example of the kind of problem that arises in this area, general management do not have a detailed understanding of the role of, say, a control systems engineer and an instrument technician, and might allocate an instrument technician to a project team to manage the production of hardware and software systems for the control of the process plant. The writer has first hand experience of this situation and knows that it does lead to failure.

Myths and popular mistakes

Project team member's beliefs

For any project to succeed it is important that the project team all have the same set of beliefs with regard to the project and it is worth reviewing and discussing these in regular project review meetings to ensure that misunderstandings and myths are dealt with promptly and do not lead to a situation where they cost time and money. The simplest way to deal with this issue is to list some of the more popular held beliefs and explain the reality, whether this applies to the client project team, the main contractor's project team or the subcontractor's project team, the following are widely held and incorrect beliefs. The contract between the two parties is what is written in the contract document. This is wrong because immediately the project begins, written agreements are usually reached between the two parties which vary the nature of the contract. This often happens because the contract when literally interpreted does not make sense or is unworkable for practical reasons. There is nothing wrong with this but it is important to understand that agreements to vary the contract are legally binding on both parties.

'I only said it, I did not confirm it in writing, therefore it is not legally binding'. This is wrong because a verbal contract is enforceable if it is witnessed. Again the two parties can vary the scope of the contract, verbally, if it is witnessed. Such verbal agreements are legally binding.

'If I do not answer the subcontractor's question and do not provide him with the information he needs, then he is to proceed on the basis of the last instruction received and because he quoted for a complete system if it does not work then it is his fault and not mine'. This is incorrect and to illustrate the point it is perhaps better to quote a practical example. The subcontractor manufacturing the motor control centre suite needs to know at a certain stage of his detailed design whether he is to design in a variable speed drive for the rotating scrapers or whether they are to be fitted with a mechanical variable speed drive. The main contractor will not know this until he has placed the subcontract for the rotating machinery with vendor A or vendor B. The dates for the exchange of information have to be agreed at the stage of entering into the subcontract, but they are binding on both parties and if the main contractor fails to provide the key information to the MCC manufacturer at the agreed time, he will be liable for the cost and time implications associated with that failure.

'If I make a mistake, an error of judgement, design error, construction error, whatever, it is better in nine out of 10 cases to brush it under the carpet rather than flag it up at an early stage and deal with it'. This is wrong, simply because if a project is being managed even moderately well, because of the reporting systems that exist it is likely that this mistake will surface somewhere and it is in the nature of projects that there is usually some sort of knock-on effect. It is far better to flag up the problem as early as possible and deal with the consequences. Bear in mind that in all fields of human activity, people make mistakes. In project management in particular, because of the pressures of time, all of the individuals working on a project tend to have to make important decisions most of which have time and cost implications, far more so than in most general management situations and in these circumstances quite logically on a large and complex project all of the project team members are going to make a fair number of mistakes (creating an open culture can help in this area tremendously).

'So long as I make sure that the subcontractor or vendor slavishly follows the specification if the equipment arrives a little bit late it doesn't matter'. This is wrong because any delay on a project has a cost implication and in all instances where correction of a defect would delay delivery of equipment, careful thought and consideration needs to be given to the matter, in many cases it may be better to deliver the equipment and modify it later or even wait until the project is complete and modify it (depending on the nature of the defect) because the knock-on effect of delay of a low cost component can amount to hundreds of thousands of pounds. Time and cost need to be at the forefront of everybody's mind when they are working on a project and this may in some instances take precedence over quality.

How to save money on a project

This could almost be entitled '... and how not to save money', but to keep it brief there are a number of key areas which make a significant difference to the success of a project.

Most organizations engaged in project management and project engineering in the UK tend to adopt a departmental structure using a matrix of engineering resource which are provided on a full and/or part time basis to project teams. In the writer's experience this is nearly always a failure, and if project structures were organized more along team lines this would give more benefits than disadvantages to the organization and make for a far more efficient business. The obvious problems with use of matrices are commitment of the individual to the project. If an individual is working on a number of projects the degree of ownership of the project and commitment to it is somewhat diluted. The politics of a departmental organization always come into play and this detracts from the efficiency of the exercise, often resulting in project managers and departmental managers entering into unnecessary conflict over resources. The task force approach to projects is always the best by far, and tends to naturally lead to high commitment, high motivation and a very efficient, cost effective form of work.

On the question of whether to use fast-track or normal approach to implementing a project, this depends very much on the nature of the project and payback. Obviously when building a gold mine, one tends to use a fast-track approach. When building a water treatment plant to meet a client's requirement which is some two years away, a more relaxed pace may be adopted. In the writer's experience, a reasonable amount of haste is always the best approach, and fast track approach can, if properly controlled, give rise to a high quality result in that the design and procurement work, construction and installation is done more quickly and efficiently and less errors tend to creep in as a result (obviously the converse is true if there is too much of a rush).

On the debate regarding appointing low cost or high cost people, because of the nature of capital projects work, errors by project team members can tend to be quite horrific in terms of the cost and time implications of them. My recommendation is therefore to employ well experienced (and consequently quite expensive) people in the key positions – project manager, commercial manager, planner, site manager, principle design and site engineer positions for mechanical, electrical, civil, instrument, controls, etc. The incremental cost difference on a project between employing inexperienced and therefore low cost personnel and employing experienced and therefore high cost personnel is small in relation to the overall project cost. The experienced people usually save substantial amounts of money because of mistakes that do not happen with the resultant saving in cost and time. This recommendation is based upon 25 years experience in the business and having seen on many occasions the results of both approaches to the problem.

One of the myths of how to save money on a project is by using subcontractors. This looks great on paper initially, but rarely results in a lower overall cost to the project. Any degree of subcontracting will reduce the direct amount of control the contractor and his team have over the scope of works being carried out by the subcontractor. The only real arguments for subcontracting are to make the volume of the work and management of the interfaces manageable and/or to make up for a technical expertise deficiency in one's own organization (which may be perfectly legitimate in the circumstances).

In short, it is a false economy to reduce the quality and quantity of people that you have working on a project, as at the start of a project this may look good on paper in the form of a financial report, but at the end of the project when the true cost of managing the interfaces becomes apparent (usually at or after final account stage) there can be a lot of tears.

The creation of a 'can-do' proactive culture in a project can make the world of difference with respect to safety, quality, cost and time and this is something that is worth spending time and money on. Creating this kind of culture is significantly easier with a task force than it is using a matrix approach to a project.

The creation of a proactive approach to value engineering and beating the programme is also worth spending time, effort and money on. This is best achieved at the start of the project by introducing for the project team members financial incentives which are project specific measurable, realistic and timed. These act as a great incentive to the project team and can result in significant time and cost savings on a project.

The creation of a simple contracting strategy is recommended without too many interfaces to be managed but ensure that the problem is not simply being passed over to your subcontractors. If a work package is put together for a subcontractor, involving the subcontractor in an extensive amount of subcontracting, then in a practical sense the risk of failure is increased. However, depending on the circumstances, this may be preferable if there is reason to believe the subcontractor is better able to manage the interfaces than one's own organization.

The use of simple readily understood systems and procedures can make a great difference to safety, quality, cost and time on a project (and the converse is true – complicated and unwieldy documents are rarely read, let alone understood).

Planning, quality and safety management

Many of the systems and procedures associated with quality and safety management are very akin to planning, that is to say they involve the project team in examining in detail what they intend to do on paper before they actually do it in reality. This approach, if kept simple and quick, can significantly improve performance on a project and is to be thoroughly recommended. If project quality plans, project health and safety plans, and project planning reports are kept concise and simple and are readily understood, then they can have a very positive effect on reducing cost and time as well as improving safety and quality. A full time project safety adviser and quality auditor is a very cost effective investment on a project and although this will never be seen on a project cost report or an accounts department report, such personnel, in the writer's experience, generally pay for themselves several times over in terms of the mistakes that are avoided.

What to do when it goes wrong

Establish the facts

As with the outbreak of warfare when a project begins to go wrong, the first casualty is the truth. Usually, if senior management take action it is at a very late stage when it becomes glaringly obvious that the project is extremely likely to fail. Generally, irrespective of the nature of the project and its duration, the first 20–30% of the time period of the project is a critical and crucial stage of the project and if the right things are not done in the right sequence at the right time during that period, then thereafter the exercise is not so much one of project management but of damage limitation. Every project reaches and passes the point where it is possible to make it a success and realistically what is possible thereafter is to limit the effects and make it a partial success (or partial failure depending on whether you are the kind of person whose glass is half full or half empty).

The important thing to do when afflicted with the problem of trying to turn around a failed project is to first establish the facts and not to be afraid if the situation warrants stopping the project in order to correct the problems (this is rarely practical for political reasons). If one does not have the luxury of stopping the project, it is necessary to look at the important things first and in priority order these are as follows.

☐ Speak to every individual working on the project team, no matter how briefly, to gather a feeling of their understanding of the project, their role in the project, what they think is going right and what they think is going wrong. Then ensure that there is a reasonably accurate scope and overall critical path analysis for the project. If there is not a programme, even at a tertiary stage, it would be preferable to employ a planner, as quickly as possible. The programme provides a route map, without it one is lost.

☐ In parallel to sorting out the programme, it would be necessary to sort out a brief understanding of the overall scope of the works (details can be studied later, and ideally no more than 2–3 weeks should be spent on this phase at this time).

☐ Ensure that there is the right quantity and quality of personnel working on the job. In practice it will take probably 2–3 months to establish this for sure, by which time in some cases it may be too late. It is therefore necessary to make a fairly instant assessment of all of the individuals and if it is considered that any one individual is not up to it (bearing in mind that one might be wrong at this stage) there should be no hesitation in replacing them. This may sound unkind, but the implications can be dire if one hesitates and if one is wrong the consequences are likely to be less dire. If it is considered that substantial numbers of the project team have to go, it should be remembered that a knowledge of the history of the project is required so one cannot afford to lose everybody. Within the first 2 weeks one needs to produce in collaboration and consultation with the project team a project quality plan and implementation plan. That document should contain what the objectives are, safety, quality, cost and time criteria and the whole project team must sign up to it. Pull the cork out of the bottleneck of progress. Find out where the hold ups are and deal with them. Usually the problem is 75% either lack of decision-making or the wrong decisions being made. It is usually the former because most people working on projects would rather not make a decision than risk making a wrong one. In the writer's experience, in 90% of cases this is the wrong approach and quite often it is better to make a wrong decision than no decision at all.

☐ It will be necessary to establish the facts of what is going wrong and to do that one needs to treat people with openness and honesty and to adopt a 'no blame' attitude. Get rid of the people you need to get rid of very quickly and then work towards building team spirit, morale and self-esteem so that everybody is pulling in the same direction. This is a lot easier said than done.

Deal with reality

Having established the facts of what has gone wrong and what has gone right, it is necessary to deal with what has gone wrong. Recriminations and punishment for sins past may give a warm feeling but will not help the bottom line. One needs to be pragmatic, avoid witch hunts and identify what the real problems are and deal with them. If the design is not complete re-employ if necessary a design team to complete it. If the equipment has not been ordered, order it and beg and plead with subcontractors and vendors in order to achieve deliveries and meet the programme but be careful not to coerce them into making promises they cannot keep. Likewise if the project has slipped there is likely to be a backlog of work for the project team and it is important that this is prioritized – do not expect everybody to catch up on everything instantaneously. One's credibility is at stake here.

Resources and planning

After validating the programme or creating one's own programme for the project, a resource plan should be prepared to suit the programme. If the project has slipped behind programme, there is likely to be a backlog of work and more resources or better quality resources may be needed to bring it back on track. The temptation to flood the project with people should be resisted. A small highly motivated but well managed team is far more effective than one which is large and unwieldy.

How to have a successful project/projects

Multiple and single projects

Principles of managing multiple projects concurrently or single projects are much the same and the temptation to stretch resources between two projects in parallel at different sites should, in the writer's view, be resisted at all cost as this is a false economy. Multiple projects must be managed by a general manager to whom the project managers of multiple projects report. Other than this, the approach to the individual projects should be identical to that described in this book in order to be successful. The idea of using a matrix resource with people working on several projects simultaneously never leads to an overall cost saving and always leads to poor safety and quality standards, prolongation of the time period for the project and increased costs.

In summary

Most experienced main contractors/subcontractors reading this document will initially react by falling about laughing, believing that it was obviously written by some idiot with a degree in psychology who has never lived in the real world. This particular idiot worked on site on his first project at the age of 14, had a father who was foolish enough to be in the same line of work, and at the age of 42, having spent a lifetime working on engineering projects, has now made most of the mistakes that anyone can possibly imagine would occur on a project, several times over. Believe me, the issues that are raised here are important for a successful project.

The potential benefits of doing the right things at the right time in the right way

By following most of the above rules, and if one is fortunate enough to achieve the 'can-do' project culture early enough in the project, the quality of the final design, time taken to implement the project and the cost of the project can all be dramatically improved because people will take it as a personal challenge to find better ways of doing the job. With the number of people involved in even a medium-sized project these days, many are going to come up with good ideas.

Consequences of failing to do the right things at the right time in the right way

In other words, project failure. That can be quite dramatic. Remember the £100 million hospital that ended up costing £200 million, or mention words such as Nimrod and Concorde to my generation, and it should send a shudder up the spine. Although project management is no 'black art', and has a lot of similarities to general management, the most fundamental difference is that in a general management situation, one usually has more of the luxury of time than one has in a project management situation, and it is more so the case that there is no substitute for appropriate experience. An experienced and hence competent project team can make the world of difference to the outcome of the project.